Memoirs of the American Mathematical Society
Number 138

Joseph A. Wolf

The action of a real semisimple Lie group on a complex flag manifold, II:

Unitary Representations on Partially Holomorphic Cohomology Spaces

Published by the
American Mathematical Society
Providence, Rhode Island
1974

Research partially supported by
N.S.F. grants GP-8008 and GP-16651, and
the Miller Institute for Basic Research in Science.

Received by the Editors March 3, 1972 and, in revised
form, February 26, 1973.

Contents

§0. Introduction.

In [49] we decomposed a complex flag manifold under the action of a real Lie group. At the end of the introduction there, we described a program for using the real group orbits as the setting for geometric realization of unitary representations of semisimple groups. Here we carry out that program for reductive Lie groups, obtaining geometric realizations for the family of representations that is involved in the Plancherel formula.

The linear semisimple case is sketched in [50].

I circulated an earlier version of this Memoir in January 1972. A number of modifications have simplified the treatment and sharpened the results here. Some of these improvements are reflected in talks ([52], [53]) given at the A.M.S. 1972 Summer Institute "Harmonic Analysis on Homogeneous Spaces."

0.1. Let G be a reductive Lie group. If $g \in G$ we assume that $\mathrm{ad}(g)$ is an inner automorphism on the complexified Lie algebra $\mathfrak{g}_C$. We also assume that G has a closed normal abelian subgroup Z that (i) centralizes the identity component G^0, (ii) has ZG^0 of finite index in G, and (iii) has $Z \cap G^0$ co-compact in the center Z_{G^0} of G^0. Let $\hat{G}$ denote the set of all equivalence classes of irreducible unitary representations of G. We work out geometric realizations for all classes in $\hat{G}$ except for a set of Plancherel measure zero.

The first part of the main body of this work is the construction and analysis of the representations for which we later obtain geometric realizations. Roughly speaking, this is a matter of extending the work of Harish-Chandra from the case where G/G^0 is finite and $[G^0, G^0]$ has finite center. We construct a series of unitary representation classes corresponding to each conjugacy class of Cartan subgroups, we compute the characters of these representations, and we use those characters for a Plancherel formula. This material is in §§2 through 5.

The second part of the main body of this work is the geometric realization of the representations just constructed. Let H represent a conjugacy class of Cartan subgroups of G. We construct partially complex homogeneous spaces Y of G, partially holomorphic G-homogeneous vector bundles $\mathcal{V} \to Y$, and Hilbert spaces $H_2^{0,q}(\mathcal{V})$ of square integrable partially harmonic (0, q)-forms on Y with values in $\mathcal{V}$. Each class $[\pi] \in \hat{G}$ corresponding to the conjugacy class of H is realized in a representation of G by its natural action on one of the $H_2^{0,q}(\mathcal{V})$. If G is compact and connected this reduces to the Bott-Borel-Weil Theorem. If H is compact this includes the discrete series realizations of Schmid and of Narasimhan-Okamoto. This material is in §1 and §§6 through 8.

In <u>§1</u> we carry out our geometric realization procedure for the "principal series" of unitary representation classes of G. This illustrates the procedure in a setting that minimizes technical problems and, consequently, gives a result which is a bit sharper than that of the general case. It is based on an analog $G = KAN$ of the Iwasawa decomposition in which $Z \subset K$. Let M be the centralizer of A in K, so $B = MAN$ is the "minimal parabolic subgroup" of G. We construct certain subgroups $U \subset M$ that contain Cartan subgroups of M, partially complex homogeneous spaces $Y = G/UAN$, and equivariant fibrations $Y \to Y/M = G/B$ whose fibres are the maximal complex subvarieties of Y. Given an irreducible unitary class $[\mu] \in \hat{U}$ and a linear functional $\sigma \in \mathfrak{a}^*$ we construct a G-homogeneous bundle $\mathcal{V}_{\mu,\sigma} \to Y$ that is holomorphic over the fibres of $Y \to Y/M$. We extend the Bott-Borel-

Weil Theorem to groups compact modulo their center, thus identifying the representations η_μ^q of M on the holomorphic sheaf cohomology spaces $H^q(M/U, \underline{\underline{O}}(\mathcal{V}_{\mu,\sigma}|_{M/U}))$. Then, given an irreducible unitary class $[\eta] \in \hat{M}$ we can find (all) classes $[\mu] \in \hat{U}$ and integers $q \geqslant 0$ with $\eta_\mu^q \in [\eta]$. That done, we prove that the "principal series" class

$$[\pi_{\eta,\sigma}] = [\mathrm{Ind}_{B\uparrow G}(\eta \otimes e^{i\sigma})], \ (\eta \otimes e^{i\sigma})(man) = e^{i\sigma}(a)\eta(m),$$

is realized as the representation of G on a certain square integrable partially holomorphic cohomology space $H_2^{0,q}(\mathcal{V}_{\mu,\sigma})$.

In §1, the representations of U and M are produced by the Peter-Weyl Theorem and characterized by Cartan's highest weight theory. In the general case we substitute an extension of Harish-Chandra's theory of the discrete series. In §1, identification of the representations η_μ^q of M comes out of a mild extension of the Bott-Borel-Weil Theorem, because the Plancherel Theorem for M is straightforward and the fibres of $Y \to Y/M$ are compact. In the general case we are forced to study these matters with some care. In §1 the induced representation process is rather easy. In the general case we need some detailed information on the distribution characters of representations induced from cuspidal parabolic subgroups.

0.2. Here is a more detailed description of the contents of this Memoir.

§1 is described above.

In §2 we write down the basic facts concerning square integrable representations of a locally compact unimodular group G relative to a unitary character $\zeta \in \hat{Z}$ where Z is a closed normal abelian subgroup. Denote

$$L_2(G/Z,\zeta) = \{f\colon G \to C \text{ measurable: } f(gz) = \zeta(z)^{-1}f(g) \text{ and } \int_{G/Z} |f(g)|^2 d(gZ) < \infty\}$$

and

$$\hat{G}_\zeta = \{[\pi] \in \hat{G}: \pi(gz) = \zeta(z)\pi(g)\}.$$

Then $\hat{G} = \bigcup_{\hat{Z}} \hat{G}_\zeta$, $L_2(G) = \int_{\hat{Z}} L_2(G/Z, \zeta)d\zeta$, and $L_2(G/Z, \zeta) = \int_{\hat{G}_\zeta} H_\pi \otimes H_\pi^* d_\zeta[\pi]$ where $d_\zeta[\pi]$ is the Plancherel measure on $\hat{G}_\zeta$. A class $[\pi] \in \hat{G}$ is called "ζ-discrete" if π is a subrepresentation of the left regular representation of G on $L_2(G/Z, \zeta)$. The set of all ζ-discrete classes forms the "ζ-discrete series" $\hat{G}_{\zeta\text{-disc}} \subset \hat{G}_\zeta$ of G. By "relative discrete series" of G we mean the union $\hat{G}_{\text{disc}} = \bigcup_{\zeta \in \hat{Z}} \hat{G}_{\zeta\text{-disc}}$. If Z is compact then $\hat{G}_{\text{disc}}$ is the usual discrete series of G.

When Z is central in G a class $[\pi] \in \hat{G}$ is in $\hat{G}_{\zeta\text{-disc}}$ if, and only if, $[\pi] \in \hat{G}_\zeta$ and the coefficient functions $\phi_{\xi,\eta}(g) = \langle \xi, \pi(g)\eta \rangle$ are in $L_2(G/Z, \zeta)$. In the general case, $\hat{G}_{\zeta\text{-disc}}$ consists of those $[\pi] \in \hat{G}_\zeta$ whose coefficient functions satisfy $|\phi_{\xi,\eta}| \in L_2(G/Z)$.

Let U be a closed subgroup of G with Z central in U and U/Z compact. In §2.4 we give a short proof that $\hat{U} = \hat{U}_{\text{disc}}$ and that every class in $\hat{U}$ is finite dimensional, and we write down the Plancherel formula for U. This is used later when we consider partially holomorphic vector bundles. Then in §2.5 we prove an exact version of Frobenius' Reciprocity Theorem relating $\hat{U}$ and $\hat{G}_{\text{disc}}$.

In <u>§3</u> we extend Harish-Chandra's theory of the discrete series to our working class of reductive Lie groups G. Given $\zeta \in \hat{Z}$ we have a central extension $1 \to S \to G[\zeta] \to ZG^0/Z \to 1$ where $S = \{s \in C: |s| = 1\}$ circle group and

$$G[\zeta] = \{S \times ZG^0\}/\{(\zeta(z)^{-1}, z): z \in Z\}.$$

We work out a bijection $G[\zeta]^\wedge_1 \to (ZG^0)^\wedge_\zeta$ where $1 \in \hat{S}$ is given by $1(s) = s$. This bijection is sufficiently explicit so that we can follow infinitesimal and distribution characters of the representations, and it restricts to a bijection $G[\zeta]^{\wedge}_{1\text{-disc}} \to (ZG^0)^{\wedge}_{\zeta\text{-disc}}$. Since $G[\zeta]$ is a connected reductive Lie group with compact center, it turns out that Harish-Chandra's discrete series theory is directly applicable to $G[\zeta]$. The bijection carries Harish-Chandra's theory over to ZG^0. Passage to G is straightforward. In particular, we see that $\hat{G}_{\text{disc}}$ is non-empty precisely when G/Z has a compact Cartan subgroup. In that case we enumerate the relative discrete series and specify the associated infinitesimal and distribution characters.

In §4 we construct a series of unitary representations of G for every conjugacy class of Cartan subgroups $H \subset G$. We refer to the series for $\{gHg^{-1}\}_{g\in G}$ as the "H-series" of G. If H/Z is compact then the H-series is the relative discrete series. If H/Z is maximally noncompact then the H-series is the principal series. If G/G^0 and the center of $[G^0, G^0]$ are finite, the various H-series are just the series constructed by Harish-Chandra and used by him to decompose $L_2(G)$. Our constructions and results are straightforward extensions of those of Harish-Chandra, Hirai and Lipsman.

Let H be a Cartan subgroup of G. We construct a "Cartan involution" θ of G that leaves H stable. Thus $\theta^2 = 1$, $\theta(H) = H$, $K = \{g \in G: \theta(g) = g\}$ contains Z, and K/Z is a maximal compact subgroup of G/Z. Now $H = T \times A$ where $T = H \cap K$ and A is the analytic subgroup for $\mathfrak{a} = \{\alpha \in \mathfrak{h}: \theta(\alpha) = -\alpha\}$. Choose a positive $\mathfrak{a}$-root system $\Sigma^+_{\mathfrak{a}}$ on $\mathfrak{g}$, define $\mathfrak{n} = \sum_{\phi\in\Sigma^+_{\mathfrak{a}}} \mathfrak{g}^{-\phi}$, and let N be the analytic subgroup of G for $\mathfrak{n}$. Then we have

$$P = \{g \in G: ad(g)N = N\} \quad \text{cuspidal parabolic subgroup of } G.$$

The centralizer of A in G has form $M \times A$ where $\theta(M) = M$; $P = MAN$, N is the unipotent radical of P, and MA is the reductive part of P. The groups M and MA inherit our working conditions from G. M has Cartan subgroup $T \supset Z$ with T/Z compact, so $\hat{M}_{disc}$ is non-empty. Let $[\eta] \in \hat{M}$ and $\sigma \in \mathfrak{a}^*$. Then

$$(\eta \otimes e^{i\sigma})(man) = e^{i\sigma}(a)\eta(m)$$

defines an irreducible unitary representation of P. The H-series of G consists of the unitary equivalence classes $[\pi_{\eta,\sigma}] = [\mathrm{Ind}_{P\uparrow G}(\eta \otimes e^{i\sigma})]$ where $[\eta] \in \hat{M}_{disc}$ and $\sigma \in \mathfrak{a}^*$.

We compute the central, infinitesimal and distribution characters of the classes $[\mathrm{Ind}_{P\uparrow G}(\eta \otimes e^{i\sigma})]$, in particular of the H-series classes of G, showing in the process that all these classes are finite sums of irreducible classes. It follows that the H-series of G depends only on the conjugacy class of H, not on the choice $\Sigma^+_{\mathfrak{a}}$ in the construction of P. It also follows that if H and H' are non-conjugate Cartan subgroups of G then every H-series class is disjoint from every H'-series class.

In <u>§5</u> we work out a Plancherel formula for G. Fix a Cartan involution θ and a system $\{H_1,\dots,H_\ell\}$ of θ-stable representatives of the conjugacy classes of Cartan subgroups of G. Let $P_j = M_jA_jN_j$ be a cuspidal parabolic subgroup of G for H_j. Fix $\zeta \in \hat{Z}$. Then in §§3 and 4, $(M_j)^\wedge_{\zeta\text{-disc}}$ was parameterized as all $[\eta_{\chi,\nu}]$ where $[\chi] \in Z_{M_j}(M_j^0)^\wedge_\zeta$, ν is in a certain discrete subset $L''_{j,\zeta}$ of $i\mathfrak{t}_j^*$, and there is a consistency condition between χ and ν. Thus the H_j-series of G is parameterized as all $[\pi_{\chi,\nu,\sigma}] = [\mathrm{Ind}_{P_j\uparrow G}(\eta_{\chi,\nu} \otimes e^{i\sigma})]$ with $\sigma \in \mathfrak{a}_j^*$. Let $\Theta_{\pi_{\chi,\nu,\sigma}}$ denote the distribution character of $[\pi_{\chi,\nu,\sigma}]$. Given $\nu \in L''_{j,\zeta}$ there are only a finite number of possibilities for $[\chi]$, and each is finite dimensional, so we have finite sums

$$\Theta_{j,\zeta,\nu+i\sigma} = \sum (\dim \chi)\Theta_{\pi_{\chi,\nu,\sigma}}.$$

We prove that there are unique (up to sets of Lebesgue measure zero) Borel functions $m_{j,\zeta,\nu}$ on $\mathfrak{a}_j^*$, $1 \leq j \leq \ell$ and $\nu \in L''_{j,\zeta}$, equivariant for the action of the Weyl group W_{G,H_j}, with the following property. Let $f \in L_2(G/Z, \zeta)$ be C^∞ with support compact modulo Z. If $x \in G$ define $r_x(f)$ by $[r_x f](g) = f(gx)$. Then

$$f(x) = \sum_{1 \leq j \leq \ell} \sum_{\nu \in L''_{j,\zeta}} |\tilde{\omega}_{\mathfrak{t}_j}(\nu)| \int_{\mathfrak{a}_j^*} \Theta_{j,\zeta,\nu+i\sigma}(r_x f) m_{j,\zeta,\nu}(\sigma) d\sigma$$

where $\tilde{\omega}_{\mathfrak{t}_j}$ is a certain function on $i\mathfrak{t}_j^*$ such that $[\eta_{\chi,\nu}]$ has formal degree $|M_j/Z_{M_j}(M_j^0)M_j^0| \cdot \dim(\chi) \cdot |\tilde{\omega}_{\mathfrak{t}_j}(\nu)|$. That is our Plancherel formula. In particular, it shows that the various H_j-classes which transform by ζ, support Plancherel measure in $\hat{G}_\zeta$.

If G/G^0 and the center of $[G^0, G^0]$ are finite, then our formula coincides with Harish-Chandra's Plancherel formula ([25],[26]). Our proof is a rather technical reduction to the case to which Harish-Chandra's results apply. In contrast to the situation of §4, this is quite non-trivial, and it requires development of some new ideas.

In §6 we describe the action of G on complex flag manifolds $X = \bar{G}_C/Q$. Here $\bar{G} = G/Z_G(G^0)$, image of G in the adjoint representation. $\bar{G}_C = \mathrm{Int}(\mathfrak{g}_C)$ is the complexification of $\bar{G}$ and Q is a parabolic subgroup of $\bar{G}_C$. This is a matter of recalling the action of $\bar{G}^0$ from [49], extending it to $\bar{G}$, and then lifting it to G. The main points here are the notion of holomorphic arc component of an orbit $G(x) \subset X$ and the analysis of measurable G-orbits. The orbits over which we realize a general H-series are measurable, and the relative discrete series classes of the definition of these H-series classes are realized over the holomorphic arc components of the orbits. At the end of §6 we give a complete classification and structure theory for a family of orbits $G(x) \subset X$ which plays a key role in the geometric realizations.

In §7 we work out the geometric realizations of the relative discrete series representation classes of G. Suppose that G has relative discrete series representations, i.e. that G/Z has a compact Cartan subgroup H/Z. Then the complex flag manifold orbits $Y = G(x) \subset \bar{G}_C/Q = X$ such that $H \subset U = \{g \in G: g(x) = x\}$ and U/Z is compact, all are open. They are obtained by choice of a positive $\mathfrak{h}_C$-root system Σ^+ on $\mathfrak{g}_C$ and an arbitrary set Φ of compact simple roots; Φ is the $\mathfrak{h}_C$-root system of $\mathfrak{u}_C$. If $[\mu] \in \hat{U}$ there is an associated G-homogeneous hermitian holomorphic vector bundle $\mathcal{V}_\mu \to Y$. For each integer $q \geqslant 0$, this specifies the Hilbert space $H_2^{0,q}(\mathcal{V}_\mu)$ of square integrable harmonic $(0,q)$-forms on Y with values in $\mathcal{V}_\mu$. G acts on $H_2^{0,q}(\mathcal{V}_\mu)$ by a unitary representation π_μ^q. We write $\Theta^{disc}_{\pi_\mu^q}$ for the sum of the distribution characters of the irreducible subrepresentations of π_μ^q.

Let $\zeta \in \hat{Z}$, $[\pi] \in \hat{G}_{\zeta\text{-disc}}$, and Θ_π the distribution character of $[\pi]$. Let ρ denote half the sum of the positive roots. Then $[\pi] = [\pi_{\chi,\lambda+\rho}]$ in the notation of §3 where $[\chi] \in Z_G(G^0)^\wedge_\zeta$ and $\lambda + \rho \in i\mathfrak{h}^*$ integrates to a unitary character on H^0 which agrees with χ on Z_{G^0}. We do this in such a way that λ is $\mathfrak{u}$-dominant, i.e. $\langle\lambda,\phi\rangle \geqslant 0$ for all $\phi \in \Phi$. Let $q(\lambda+\rho)$ denote the number of compact positive roots with $\langle\lambda+\rho, \phi\rangle < 0$ plus the number of noncompact positive roots with $\langle\lambda+\rho, \phi\rangle > 0$. Let $[\mu^0] \in \hat{U}^0$ be the class with highest weight λ; $U = Z_G(G^0)U^0$ so $[\mu] = [\chi \otimes \mu^0] \in \hat{U}$. We prove the following.

1. $\displaystyle\sum_{q\geqslant 0} (-1)^q \, \Theta^{disc}_{\pi_\mu^q} = (-1)^{|\Sigma^+|+q(\lambda+\rho)} \, \Theta_{\pi_{\chi,\lambda+\rho}}.$

2. There is a constant $b \geqslant 0$ depending only on $[\mathfrak{g},\mathfrak{g}]$ such that, if

$|\langle \lambda+\rho, \phi\rangle| > b$ for every $\phi \in \Sigma^+$ and $q \neq q(\lambda+\rho)$, then $H_2^{0,q}(\mathcal{V}_\mu) = 0$.

3. If $q_0 \geqslant 0$ is an integer, and if $q \neq q_0$ implies $H_2^{0,q}(\mathcal{V}_\mu) = 0$, then $[\pi_\mu^{q_0}] = [\pi_{\chi,\lambda+\rho}] = [\pi] \in \hat{G}_{\zeta\text{-disc}}$.

This gives explicit geometric realizations for "most" of the relative discrete series classes in $\hat{G}$, and it gives implicit geometric realizations for all of $\hat{G}_{disc}$.

The idea of working with an alternating sum formula and a vanishing theorem goes back in rough form to Bott's proof of the Bott-Borel-Weil Theorem [4], and in precise form to the Lie algebra cohomology proof given by Kostant [29]. It is used in the Narasimhan-Okamoto treatment [38] of the case where G is a connected linear semisimple group and Y is an hermitian symmetric space. The vanishing theorem (2) goes back to Griffiths-Schmid [9]. Our proof of these results uses methods developed by Schmid [43], who proved (2) and (3) in case G is a connected semisimple group with finite center and $Y = G/H$. But there are important differences, for Schmid could fall back on Harish-Chandra's Plancherel formula, while we have to develop a Plancherel formula valid for our class of groups.

In §8 we work out the geometric realizations for all the nondegenerate series classes of unitary representations of G.

Fix a Cartan subgroup $H = T \times A$ of G and a corresponding cuspidal parabolic subgroup $P = MAN$ of G. The H-series classes are realized over measurable orbits $Y = G(x) \subset \bar{G}_C/Q = X$ such that (i) the G-normalizer $N_{[x]}$ of the holomorphic arc component $S_{[x]}$ through x is an open subgroup of P and (ii) $T \subset U = \{m \in M: m(x) = x\}$ with U/Z compact. Such pairs (X, x) always exist, and we classified a family of them at the end of §6. Then G has isotropy subgroup UAN at x, $N_{[x]} = M^\dagger AN$ where $M^\dagger = Z_M(M^0)M^0$, and

$U = Z_M(M^0)U^0$ with $U \cap M^0 = U^0$. Let $\rho_{\mathfrak{a}} = \frac{1}{2}\sum_{\phi\in\Sigma_{\mathfrak{a}}^+} (\dim \mathfrak{g}^{\phi})\phi$ where $\mathfrak{n}$ is the sum of the negative $\mathfrak{a}$-root spaces. If $[\mu] \in \hat{U}$ and $\sigma \in \mathfrak{a}^*$ we have a G-homogeneous vector bundle $\mathcal{V}_{\mu,\sigma} \to G/UAN = Y$ associated to the representation $\mu \otimes e^{\rho_{\mathfrak{a}}+i\sigma}$ of UAN. This bundle carries a K-invariant hermitian metric, and it is holomorphic over every holomorphic arc component of Y. This leads us to the Hilbert spaces $H_2^{0,q}(\mathcal{V}_{\mu,\sigma})$ of all square integrable partially harmonic (0,q)-forms on Y with values in $\mathcal{V}_{\mu,\sigma}$. G acts on $H_2^{0,q}(\mathcal{V}_{\mu,\sigma})$ by a unitary representation $\pi_{\mu,\sigma}^q$. As first step toward their identification, we prove $[\pi_{\mu,\sigma}^q] = [\mathrm{Ind}_{N_{[x]}}{\uparrow}G(\eta_\mu^q \otimes e^{i\sigma})]$ where η_μ^q is the representation of $M^\dagger$ on $H_2^{0,q}(\mathcal{V}_{\mu,\sigma}|S_{[x]})$.

Decompose $[\mu] = [\chi \otimes \mu^0]$ where $[\chi] \in Z_M(M^0)^\wedge$ and $[\mu^0] \in \hat{U}^0$ has highest weight ν; then $\chi = e^\nu$ on the center of M^0. Suppose that $\nu + \rho_{\mathfrak{t}}$ is $\mathfrak{m}$-regular. We prove the following.

1. The H-series constituents of $\pi_{\mu,\sigma}^q$ are just its irreducible subrepresentations, and their sum ${}^H\pi_{\mu,\sigma}^q$ has well defined distribution character $\Theta^H_{\pi_{\mu,\sigma}^q}$.

2. $$\sum_{q\geqslant 0} (-1)^q \Theta^H_{\pi_{\mu,\sigma}^q} = (-1)^{|\Sigma_{\mathfrak{t}}^+|+q_M(\nu+\rho_{\mathfrak{t}})}\, \Theta_{\pi_{\chi,\nu+\rho_{\mathfrak{t}},\sigma}}$$

where $[\pi_{\chi,\nu+\rho_{\mathfrak{t}},\sigma}]$ is the H-series class so denoted in §4.

3. There is a constant $b_H \geqslant 0$ that depends only on $[\mathfrak{m},\mathfrak{m}]$, such that if $|\langle \nu+\rho_{\mathfrak{t}}, \phi\rangle| > b_H$ for all $\phi \in \Sigma_{\mathfrak{t}}^+$ then $H_2^{0,q}(\mathcal{V}_{\mu,\sigma}) = 0$ for all integers $q \neq q_M(\nu+\rho_{\mathfrak{t}})$.

4. If q_0 is an integer such that $q \neq q_0$ implies $H_2^{0,q}(\mathcal{V}_{\mu,\sigma}) = 0$, then $\pi_{\mu,\sigma}^{q_0}$ belongs to the H-series class $[\pi_{\chi,\nu+\rho_{\mathfrak{t}},\sigma}]$.

This gives explicit geometric realizations for "most" of the H-series classes of unitary representations of G, and it gives implicit geometric realizations for all of them. Thus, for every $\zeta \in \hat{Z}$ we have geometric realizations of a subset of $\hat{G}_\zeta$ that supports Plancherel measure there.

0.3. There are a number of convenient classes of groups for which one can carry out our geometric realization procedure. In this Memoir the procedure is worked through for reductive Lie groups G such that if $g \in G$ then $ad(g)$ is an inner automorphism of $\mathfrak{g}_C$ and G has a closed normal abelian subgroup Z with the properties

(0.3.1a) Z centralizes the identity component G^0,

(0.3.1b) ZG^0 has finite index in G, and

(0.3.1c) $Z \cap G^0$ is co-compact in the center Z_{G^0} of G^0.

The differences between that class and some others are worth discussing. Here we always assume that G^0 is a reductive Lie group, but G/G^0 might be non-discrete.

In setting up the harmonic theory for geometric realizations, one needs to stay with finite dimensional bundles. Thus he is forced to make assumptions on the structure of G modulo G^0 to ensure that finite dimensionality. In order of increasing severity, the natural sets of hypotheses are

(0.3.2a) G has a closed normal abelian subgroup co-compact in $Z_G(G^0)$;

(0.3.2b) G has a closed normal abelian subgroup that satisfies (0.3.1);

(0.3.2c) the component group G/G^0 is finite; and

(0.3.2d) G is connected.

Again in increasing order of severity, natural sets of hypotheses for the action of G on G^0 are

(0.3.3a) no condition;

(0.3.3b) $ad(G)$ is compact modulo inner automorphisms of G^0;

(0.3.3c) ad(G) is finite modulo inner automorphisms of G^0;

(0.3.3d) if $g \in G$ then ad(g) is an inner automorphism of $\mathfrak{g}_C$; and

(0.3.3e) if $g \in G$ then ad(g) is an inner automorphism of G^0.

Finally, the natural assumptions on the structure of the connected reductive Lie group G^0 are

(0.3.4a) no condition;

(0.3.4b) the derived group $[G^0, G^0]$ is closed in G^0;

(0.3.4c) the derived group $[G^0, G^0]$ has finite center; and

(0.3.4d) G^0 has finite center.

It is convenient to work under a set of conditions that is hereditary in the sense: if $P = MAN$ cuspidal parabolic subgroup of G, then M inherits the conditions from G.

Condition (0.3.2a) is hereditary. It implies (0.3.3b) and gives a workable class of unimodular locally compact groups. Compare Lipsman [31] and extend his work in the obvious way to $Z_G(G^0)$ by the method of §2.4 below. I have not written this down because I do not know any interesting examples that are not covered by (0.3.2b) = (0.3.1).

Condition (0.3.2b) is hereditary. It implies (0.3.3c) and gives a nice class of unimodular locally compact groups. The reader will see that our further assumption that G be a Lie group is of no importance; given $\zeta \in \hat{Z}$ the central extension $G[\zeta]$ of ZG^0 in §3.3 is automatically a connected reductive Lie group.

Condition (0.3.2c) is not hereditary, but (0.3.2c) together with (0.3.4c) is a hereditary set of conditions. Real reductive linear algebraic groups satisfy (0.3.2c) and (0.3.4c). Much of Harish-Chandra's work on harmonic analysis for real Lie groups is done under the hypotheses (0.3.2c) and (0.3.4c). I have avoided (0.3.4c) because it excludes the universal covering groups of the automorphism groups of the bounded symmetric domains; those are important semisimple groups.

Condition (0.3.2d) cannot be made hereditary without requiring G to be an abelian extension of a compact group.

Condition (0.3.3d) is hereditary. Harish-Chandra used it in extending his theory of the discrete series to disconnected groups [26]. We use it for several reasons: the relative discrete series and the other nondegenerate series are much more easily described, every primary representation has an infinitesimal character, there is no problem about whether G acts on a complex flag manifold $X = \mathrm{Int}(\mathfrak{g}_C)/Q$, and the G-orbits on X are easier to understand.

One can weaken (0.3.3d) to (0.3.3c), without serious mathematical problems, by inducing all representations from $\{g \in G\colon \mathrm{ad}(g)$ is an inner automorphism of $\mathfrak{g}_C\}$ to G. But then the theorems become somewhat complicated.

Condition (0.3.3e) cannot be made hereditary.

Conditions (0.3.4b) and (0.3.4d) cannot be made hereditary. Our reason for avoiding (0.3.4c) is mentioned above.

In summary, we work with the smallest convenient hereditary class that contains all connected semisimple real Lie groups.

0.4. It is a pleasure to acknowledge a number of helpful communications and conversations with Harish-Chandra, Calvin C. Moore and Marc Rieffel. Harish-Chandra clarified my understanding of his work. Calvin Moore told me the basic facts of harmonic analysis. Marc Rieffel suggested direct construction of the central extensions $G[\zeta]$ of §3 rather than their construction through the Bargmann-Mackey-Moore theory ([3]; [34], [35]; [2], [36], [37]) of cocycle representations.

§1. The Principal Series.

We illustrate our geometric realization procedure by carrying it out for the "principal series" of unitary representation classes of reductive Lie groups. This is a setting that minimizes technical problems and yields a result which is a bit sharper than that of the general situation. We first consider groups compact modulo their centers and extend the Bott-Borel-Weil Theorem to those groups. We then define the principal series, describe the geometric setting, and formulate and prove our geometric realization theorem. The role of the extended Bott-Borel-Weil Theorem in our principal series realizations, illustrates the role of discrete series realizations (§8) in our geometric realizations of nondegenerate series classes (§9).

1.1. Let M be a reductive Lie group. In other words, its Lie algebra $\mathfrak{m}$ is direct sum of an abelian ideal (the center) and a semisimple ideal $[\mathfrak{m}, \mathfrak{m}]$. We assume

(1.1.1) if $m \in M$ then $ad(m)$ is an inner automorphism on $\mathfrak{m}_C$.

We also assume that M has a closed normal abelian subgroup Z such that

(1.1.2a) Z centralizes the identity component M^0 of M and

(1.1.2b) M/Z is compact.

From (1.1.2), ZM^0 has finite index in M and $Z \cap M^0$ is co-compact in the center Z_{M^0} of M^0.

If η is a (continuous) unitary representation of M then $[\eta]$ denotes its unitary equivalence class. The set of all equivalence classes of irreducible unitary representations of M is denoted $\hat{M}$. If E is a closed central subgroup of M and $\xi \in \hat{E}$ then $\hat{M}_\xi$ denotes $\{[\eta] \in \hat{M}: \eta|_E$ is a multiple of $\xi\}$.

We collect some facts on representations of M.

1.1.3. Proposition. *Let* $Z_M(M^0)$ *denote the* M-*centralizer of* M^0, *so* $Z_M(M^0) \cap M^0 = Z_{M^0}$.

1. $M = Z_M(M^0)M^0$.

2. *Every irreducible unitary representation of* M *is finite dimensional.*

3. *Given an irreducible unitary class* $[\eta] \in \hat{M}$, *there exist unique* $\xi \in \hat{Z}_{M^0}$, $[\chi] \in Z_M(M^0)^\wedge_\xi$ *and* $[\eta^0] \in (M^0)^\wedge_\xi$ *such that* $[\eta] = [\chi \otimes \eta^0]$.

4. *Let* $\mathfrak{t} \subset \mathfrak{m}$ *Cartan subalgebra,* $\Sigma^+_{\mathfrak{t}}$ *a positive root system,* $\rho_{\mathfrak{t}}$ *half the sum of the positive roots,* $T^0 = \exp(\mathfrak{t})$ *and*

$L^+_{\mathfrak{m}} = \{\nu \in i\mathfrak{t}^* : e^{\nu-\rho_{\mathfrak{t}}} \in \hat{T}^0$ *well defined and* $\langle \nu, \phi \rangle > 0$ *for every* $\phi \in \Sigma^+_{\mathfrak{t}}\}$. *Then there is a bijection* $\nu \to [\eta^0_\nu]$ *of* $L^+_{\mathfrak{m}}$ *onto* $\hat{M}^0$ *given by:* $\nu - \rho_{\mathfrak{t}}$ *is the highest weight of* η^0_ν. *Further* $[\eta^0_\nu] \in (M^0)^\wedge_\xi$ *where* $\xi = e^{\nu-\rho_{\mathfrak{t}}}\big|_{Z_{M^0}}$.

Proof. If $m \in M$ then (1.1.1) and (1.1.2) force $\mathrm{ad}(m)$ to be an inner automorphism on M^0. In other words, $M = Z_M(M^0)M^0$. In particular Z_{M^0} is central in M.

Let $[\eta] \in \hat{M}$. Now there is a unitary character ξ on Z_{M^0} such that $\eta|_{Z_{M^0}}$ is a multiple of ξ. Since M^0/Z_{M^0} and $Z_M(M^0)/Z_{M^0}$ are compact, the groups M^0 and $Z_M(M^0)$ are of type I with every irreducible unitary representation finite dimensional; see §2.4 below. Now $[\eta|_{Z_M(M^0)}]$ is a multiple of a finite dimensional class $[\chi] \in Z_M(M^0)^\wedge_\xi$ and $[\eta|_{M^0}]$ is a multiple of a finite dimensional class $[\eta^0] \in (M^0)^\wedge_\xi$. By irreducibility, $[\eta]$ coincides with its subrepresentation $[\chi \otimes \eta^0]$. This proves assertions 2 and 3.

The last assertion is É. Cartan's highest weight theory for representations of $\mathfrak{m}$. *q.e.d.*

We now extend the Bott-Borel-Weil Theorem to give geometric realizations of the classes in $\hat{M}$. Retain the notation of Proposition 1.1.3.

The homogeneous kaehler manifolds of M are the manifolds S_Φ constructed as follows. Denote

(1.1.4a) $\Pi_{\mathfrak{t}}$: simple $\mathfrak{t}_C$-root system on $\mathfrak{m}_C$ for $\Sigma_{\mathfrak{t}}^+$.

Now choose

(1.1.4b) Φ: arbitrary subset of $\Pi_{\mathfrak{t}}$.

This data gives us

(1.1.5a) $\mathfrak{z}_\Phi = \{x \in \mathfrak{t} : \Phi(x) = 0\}$ and $Z_\Phi^0 = \exp(\mathfrak{z}_\Phi) \subset T^0$.

Define

(1.1.5b) U_Φ is the M-centralizer of Z_Φ^0 and $S_\Phi = M/U_\Phi$.

Proposition 1.1.3(1) allows us to define

(1.1.6a) $\bar{M} = M/Z_M(M^0) = M^0/Z_{M^0}$ compact connected Lie group.

Evidently, the M-centralizer T of T^0 satisfies

(1.1.6b) $\bar{T} = T/Z_M(M^0) = T^0/Z_{M^0}$ is a maximal torus in $\bar{M}$.

Since Z_Φ^0/Z_{M^0} is a sub-torus of $\bar{T}$ now

(1.1.6c) Φ is a simple $\mathfrak{t}_C$-root system for $\mathfrak{u}_{\Phi C}$.

We use (1.1.6) to describe the kaehler structure on $S_\Phi = M/U_\Phi$. Note that $\bar{M}$ has complexification

(1.1.7a) $\bar{M}_C = \mathrm{Int}(\mathfrak{m}_C)$ inner automorphism group of $\mathfrak{m}_C$.

Now define $\bar{U}_\Phi = U_\Phi/Z_M(M^0)$. It is connected. Define

(1.1.7b) $\mathfrak{r}_\Phi = \bar{\mathfrak{u}}_{\Phi C} + \sum_{\phi \in \Sigma_{\mathfrak{t}}^+} \bar{\mathfrak{m}}^{-\phi}$ subalgebra of $\bar{\mathfrak{m}}_C$

and

(1.1.7c) R_Φ is the complex analytic subgroup of $\bar{M}_C$ for $\mathfrak{r}_\Phi$.

It is standard that R_Φ is closed in $\bar{M}_C$. Let $m \mapsto \bar{m}$ denote the composition $M \to \bar{M} \subset \bar{M}_C$. M acts on $\bar{M}_C/R_\Phi$ by this projection. We assert

1.1.8. Lemma. *The map* $mU_\Phi \to \bar{m}R_\Phi$ *is an* M-*equivariant bijection of* $S_\Phi = M/U_\Phi$ *onto* $\bar{M}_C/R_\Phi$.

Proof. The M-orbit of $1 \cdot R_\Phi$ is closed because $\bar{M}$ is compact. Since $\bar{\mathfrak{m}} \cap \mathfrak{r}_\Phi = \bar{\mathfrak{u}}_\Phi$, the isotropy subalgebra of $\mathfrak{m}$ at $1 \cdot R_\Phi$ is $\mathfrak{u}_\Phi$, and the orbit has real dimension $\dim_R \mathfrak{m} - \dim_R \mathfrak{u}_\Phi = \dim_R \bar{\mathfrak{m}} - \dim_R \bar{\mathfrak{u}}_\Phi = \dim_R \bar{\mathfrak{m}}_C - \dim_R \mathfrak{r}_\Phi$. Thus the orbit is open, hence equal to all of $\bar{M}_C/R_\Phi$, and $mU_\Phi \to \bar{m}R_\Phi$ is a covering space. Since $\bar{M}_C/R_\Phi$ is simply connected, the proof is complete. *q.e.d.*

The M-homogeneous complex structure on S_Φ is inherited from the complex presentation $S_\Phi = \bar{M}_C/R_\Phi$ by means of Lemma 1.1.8. The M-homogeneous kaehler metrics are the Lie algebra coboundaries of $\mathfrak{u}_\Phi$-regular elements of $\mathfrak{z}_\Phi^*$.

The irreducible M-homogeneous holomorphic vector bundles $\mathcal{V}_\mu \to S_\Phi$ are constructed as follows. Let

(1.1.9a) $[\mu] \in \hat{U}_\Phi$ and V_μ is its representation space.

Then we have the associated complex vector bundle

(1.1.9b) $\mathcal{V}_\mu \to M/U_\Phi = S_\Phi$ M-homogeneous, hermitian.

We assert the existence of a complex structure on $\mathcal{V}_\mu$ such that

(1.1.9c) $\mathcal{V}_\mu \to S_\Phi$ M-homogeneous holomorphic vector bundle.

This is a special case of Lemma 7.1.4 below. Here there is a direct proof because the derived group $M^{(1)} = [M^0, M^0]$ is compact, connected and semisimple. Projection $M \to \bar{M}$ restricts to a finite covering $M^{(1)} \to \bar{M}$, and that complexifies to a finite holomorphic covering $p: M_C^{(1)} \to \bar{M}_C$. Let $\mu^{(1)}$ be the restriction of μ to $M^{(1)} \cap U_\Phi$. Denote $R_\Phi^{(1)} = p^{-1}(R_\Phi)$; it is connected. Now $\mu^{(1)}$ has a unique completely reducible holomorphic extension $\mu_C^{(1)}$ to $R_\Phi^{(1)}$. That gives the structure of $M_C^{(1)}$-homogeneous holomorphic vector bundle to

$$\mathcal{V}_\mu = \mathcal{V}_{\mu_C^{(1)}} \to M_C^{(1)}/R_\Phi^{(1)} = \bar{M}_C/R_\Phi = S_\Phi.$$

This structure is stable under the action of M, so (1.1.9c) is proved.

We now have

(1.1.10a) $\underline{O}(\mathcal{V}_\mu)$: sheaf of germs of holomorphic sections of $\mathcal{V}_\mu$ and

(1.1.10b) η_μ^q: natural representation of M on $H^q(S_\Phi; \underline{O}(\mathcal{V}_\mu))$.

The representations η_μ^q become unitary when we use the Kodaira-Hodge harmonic theory to identify $H^q(S_\Phi; \underline{O}(\mathcal{V}_\mu))$ with

(1.1.10c) $H^{0,q}(\mathcal{V}_\mu)$: harmonic $(0,q)$-forms on S_Φ with values in $\mathcal{V}_\mu$.

Combining Proposition 1.1.3(1) with the consequence $U_\Phi \cap M_\Phi^0 = U_\Phi^0$ of simple connectivity of S_Φ,

(1.1.11a) $M = Z_M(M^0)M^0$, $U_\Phi = Z_M(M^0)U_\Phi^0$ and $Z_M(M^0) \cap U_\Phi^0 = Z_{M^0}$.

Now, as in Proposition 1.1.3,

(1.1.11b) $\hat{U}_\Phi = \bigcup_{\xi \in \widehat{Z_{M^0}}} (U_\Phi)^\wedge_\xi$ disjoint union, where

(1.1.11c) $(U_\Phi)^\wedge_\xi = \{[\chi \otimes \mu^0]: [\chi] \in Z_M(M^0)^\wedge_\xi \text{ and } [\mu^0] \in (U_\Phi^0)^\wedge_\xi\}$.

Finally, elements of $\hat{U}_\Phi^0$ are specified by their highest weights in the positive $\mathfrak{t}_C$-root system generated by Φ.

The Bott-Borel-Weil Theorem [4] extends to our situation as follows.

<u>1.1.12. Proposition</u>. *Let* $[\mu] \in \hat{U}_\Phi$, *say* $[\mu] = [\chi \otimes \mu^0]$ *as in* (1.1.11), *and let* β *be the highest weight of* μ^0.

1. *If* $\langle \beta + \rho_{\mathfrak{t}}, \phi \rangle = 0$ *for some* $\phi \in \Sigma_{\mathfrak{t}}^+$, *then* $H^q(S_\Phi; \underline{O}(\mathcal{V}_\mu)) = 0$ *for all integers* q.

2. *If* $\langle \beta + \rho_{\mathfrak{t}}, \phi \rangle \neq 0$ *for every* $\phi \in \Sigma_{\mathfrak{t}}^+$, *define* q_0 *to be the number of roots* $\phi \in \Sigma_{\mathfrak{t}}^+$ *for which* $\langle \beta + \rho_{\mathfrak{t}}, \phi \rangle$ *is negative, and let* ν *denote the unique element of* $L_{\mathfrak{m}}^+$ *that is conjugate to* $\beta + \rho_{\mathfrak{t}}$ *by an element of the Weyl group of* (M^0, T^0). *Then* $[\eta_\mu^{q_0}] = [\chi \otimes \eta_\nu^0]$ *and* $H^q(S_\Phi; \underline{O}(\mathcal{V}_\mu)) = 0$ *for*

every integer $q \neq q_0$.

Proof. We reduce to the (usual) Bott-Borel-Weil Theorem for compact connected semisimple groups.

From $[\mu] = [\chi \otimes \mu^0]$ the representation space $V_\mu = V_\chi \otimes V_{\mu^0}$. As $Z_M(M^0)$ acts trivially on S_Φ now the bundle $\mathcal{V}_\mu = V_\chi \otimes \mathcal{V}_{\mu^0}$. Thus $[\eta^q_\mu] = [\chi \otimes \eta^q_{\mu^0}]$ where $\eta^q_{\mu^0}$ denotes the representation of M^0 on $H^q(S_\Phi; \underline{\mathcal{O}}(\mathcal{V}_{\mu^0}))$. We have reduced to the case where M is connected.

Now that M is connected, $M = Z^0_M M^{(1)}$ where $M^{(1)} = [M,M]$ compact connected semisimple, and where $F = Z^0_M \cap M^{(1)}$ is finite. From (1.1.6c) and (1.1.11a), also $U_\Phi = Z^0_M U^{(1)}_\Phi$ where $U^{(1)}_\Phi = U_\Phi \cap M^{(1)}$, and $Z^0_M \cap U^{(1)} = F$. Split $[\mu] = [\varepsilon \otimes \mu^{(1)}]$ where $[\varepsilon] \in \hat{Z}^0_M$ and $[\mu^{(1)}] \in \hat{U}^{(1)}_\Phi$ give the same character $\varepsilon|_F$ on F. As above $[\eta^q_\mu] = [\varepsilon \otimes \eta^q_{\mu^{(1)}}]$ where $\eta^q_{\mu^{(1)}}$ denotes the representation of $M^{(1)}$ on $H^q(S_\Phi; \underline{\mathcal{O}}(\mathcal{V}_\mu))$. We have reduced to the case where M is compact, connected and semisimple. *q.e.d.*

<u>1.2</u>. Let G be a reductive Lie group. Thus its Lie algebra $\mathfrak{g} = \mathfrak{c} \oplus [\mathfrak{g},\mathfrak{g}]$ where $\mathfrak{c}$ is the center and $[\mathfrak{g},\mathfrak{g}]$ is semisimple. We assume

(1.2.1) if $g \in G$ then $\mathrm{ad}(g)$ is an inner automorphism on $\mathfrak{g}_C$.

We also assume that G has a closed normal abelian subgroup Z such that

(1.2.2a) Z centralizes the identity component G^0 of G,

(1.2.2b) ZG^0 has finite index in G, and

(1.2.2c) $Z \cap G^0$ is co-compact in the center Z_{G^0} of G^0.

As noted in §1.1, (1.1.2) is a special case of (1.2.2).

We recall the minimal parabolic subgroups of G. Choose

(1.2.3a) θ: Cartan involution of G.

In other words, θ is an involutive automorphism of G whose fixed point set

(1.2.3b) $K = \{g \in G: \theta(g) = g\}$

has the property

(1.2.3c) $Z_G(G^0) \subset K$ and $K/Z_G(G^0)$ is a maximal compact subgroup of $G/Z_G(G^0)$.

See Lemmas 4.1.1 and 4.1.2 below for the construction and conjugacy of Cartan involutions. Now choose

(1.2.4a) $\mathfrak{a}$: maximal abelian subspace of $\{x \in \mathfrak{g} : \theta(x) = -x\}$.

Any two such choices are $\mathrm{ad}_G(K)$-conjugate. Also, $\mathrm{ad}(\mathfrak{a})$ is completely reducible on $\mathfrak{g}$ with real eigenvalues. Now $\mathfrak{g}$ is direct sum of its subspaces

(1.2.4b) $\mathfrak{g}^{\psi} = \{x \in \mathfrak{g} : [a,x] = \psi(a)x$ for all $a \in \mathfrak{a}\}$ where $\psi \in \mathfrak{a}^*$.

The $\mathfrak{a}$-<u>root system</u> on $\mathfrak{g}$ is $\Sigma_{\mathfrak{a}} = \{\psi \in \mathfrak{a}^* - \{0\} : \mathfrak{g}^{\psi} \neq 0\}$. Choose

(1.2.4c) $\Sigma_{\mathfrak{a}}^+$: positive $\mathfrak{a}$-root system on $\mathfrak{g}$.

Any two such choices are conjugate by the K-normalizer of $\mathfrak{a}$. The pair $(\mathfrak{a}, \Sigma_{\mathfrak{a}}^+)$ specifies

(1.2.5a) N: analytic subgroup of G for $\mathfrak{n} = \sum_{\psi \in \Sigma_{\mathfrak{a}}^+} \mathfrak{g}^{-\psi}$.

The corresponding <u>minimal parabolic subgroup</u> of G is

(1.2.5b) B: normalizer of N in G.

Let us denote

(1.2.6a) A: analytic subgroup of G for $\mathfrak{a}$ and

(1.2.6b) M: centralizer of A in K.

Then $MA = M \times A$ is the centralizer of A in G, and $G = KAN$ corresponds to the Iwasawa decomposition. Using the corresponding standard result for the identity component of $G/Z_G(G^0)$, we see that

(1.27) $B = MAN$, and M satisfies (1.1.1) and (1.1.2).

Retain the notation of Proposition 1.1.3. There is a positive root system Σ^+ for $\mathfrak{g}_C$ relative to its Cartan subalgebra $(\mathfrak{t} + \mathfrak{a})_C$, such that

$$\Sigma^+_{\mathfrak{a}} = \{\gamma|_{\mathfrak{a}} : \gamma \in \Sigma^+ \text{ and } \gamma|_{\mathfrak{a}} \neq 0\}$$

and

$$\Sigma^+_{\mathfrak{t}} = \{\gamma|_{\mathfrak{t}} : \gamma \in \Sigma^+ \text{ and } \gamma|_{\mathfrak{a}} = 0\}.$$

Now let

(1.2.8a) $[\chi \otimes \eta^0_\nu] \in \hat{M}$ and $\sigma \in \mathfrak{a}^*$.

That specifies

(1.2.8b) $[\alpha_{\chi,\nu,\sigma}] \in \hat{B}$ by $\alpha_{\chi,\nu,\sigma}(man) = (\chi \otimes \eta^0_\nu)(m)e^{i\sigma}(a)$.

The corresponding <u>principal series representation</u> of G is

(1.2.8c) $\pi_{\chi,\nu,\sigma} = \mathrm{Ind}_{B\uparrow G}(\alpha_{\chi,\nu,\sigma})$, unitarily induced representation.

We construct partially complex homogeneous spaces Y_Φ of G, over which the principal series classes $[\pi_{\chi,\nu,\sigma}]$ will be realized. Retain (1.1.4) and denote

$$\Pi:\ \text{simple } (\mathfrak{t}+\mathfrak{a})_C\text{-root system on } \mathfrak{g}_C \text{ for } \Sigma^+.$$

Then $\Phi \subset \Pi_{\mathfrak{t}} \subset \Pi$. Now denote

(1.2.9a) $\bar{G} = G/Z_G(G^0)$, so $\bar{G}_C = \mathrm{Int}(\mathfrak{g}_C)$ and $\bar{\mathfrak{g}} = \mathfrak{g}/\mathfrak{c}$.

Using (1.1.5) we have

(1.2.9b) $\mathfrak{q}_\Phi = (\mathfrak{u}_\Phi/\mathfrak{c})_C + \sum_{\gamma\in\Sigma^+} \mathfrak{g}^{-\gamma}$ complex subalgebra of $\bar{\mathfrak{g}}_C$.

Now define

(1.2.9c) Q_Φ is the complex analytic subgroup of $\bar{G}_C$ for $\mathfrak{q}_\Phi$,

(1.2.9d) $X_\Phi = \bar{G}_C/Q_\Phi$ compact complex homogeneous space.

Then Q_Φ is its own normalizer in $\bar{G}_C$, so we can identify X_Φ with the set of all G_C-conjugates of Q_Φ. Our assumption (1.2.1) now says that

(1.2.10a) G acts on X_Φ by holomorphic transformations.

Now define

(1.2.10b) $x_\Phi = 1\cdot Q_\Phi \in X_\Phi$ and $Y_\Phi = G(x_\Phi) \subset X_\Phi$.

Then, as in the proof of Lemma 1.1.8,

(1.2.10c) $U_\Phi AN = \{g \in G: g(x_\Phi) = x_\Phi\}$ so $Y_\Phi = G/U_\Phi AN$.

In particular, Y_Φ contains

(1.2.11a) $S_\Phi = M(x_\Phi) = (MAN)(x_\Phi)$ complex submanifold of X_Φ.

Since the minimal parabolic $B = MAN$ is its own G-normalizer, one can look at the Lie algebra level to see

(1.2.11b) $B = \{g \in G: gS_\Phi = S_\Phi\}$.

We will need the facts given by

1.2.12. Lemma. *The* gS_Φ, $g \in G$, *are complex submanifolds of* X_Φ *contained in* Y_Φ. *If* $S \subset Y_\Phi$ *is a connected complex submanifold of* X_Φ *then* S *is contained in one of the* gS_Φ. *Finally* $Y_\Phi \to G/B = K/M$ *is a well defined equivariant fibration*, gS_Φ *being the fibre over* gB.

Indication of proof. In the terminology of §6.5, the assertion is that the topological component of x_Φ in S_Φ is the holomorphic arc component of Y_Φ through x_Φ. Let τ denote complex conjugation. Then $\mathfrak{q}_\Phi + \tau\mathfrak{q}_\Phi = (\mathfrak{m}/\mathfrak{c})_C + \mathfrak{a}_C + \mathfrak{n}_C = (\mathfrak{b}/\mathfrak{c})_C$ subalgebra of $\mathfrak{g}_C$, i.e. Y_Φ is integrable in the terminology of §6.6. Further $\mathfrak{q}_\Phi$ has τ-stable reductive part $(\mathfrak{u}_\Phi/\mathfrak{c})_C$. Now (1.2.11) and (6.6.6) give the assertion. *q.e.d.*

We can now construct the partially holomorphic vector bundles over Y_Φ and the partially holomorphic cohomology spaces on which we will realize the principal series of G. Let

(1.2.13a) $[\mu] \in \hat{U}_\Phi$, $\sigma \in \mathfrak{a}^*$ and $\rho_{\mathfrak{a}} = \frac{1}{2} \sum_{\phi \in \Sigma_{\mathfrak{a}}^+} (\dim \mathfrak{g}^\phi)\phi \in \mathfrak{a}^*$.

Then we have a representation $\gamma_{\mu,\sigma}$ of $U_\Phi AN$ on V_μ, given by

(1.2.13b) $\gamma_{\mu,\sigma}(uan) = e^{\rho_{\mathfrak{a}}+i\sigma}(a)\mu(u)$.

Denote the associated G-homogeneous complex vector bundle by

(1.2.13c) $\mathcal{V}_{\mu,\sigma} \to G/U_\Phi AN = Y_\Phi$; so $\mathcal{V}_{\mu,\sigma}|_{S_\Phi} = \mathcal{V}_\mu$.

Now each $\mathcal{V}_{\mu,\sigma}|_{gS_\Phi} \to gS_\Phi$ is an ad(g)B-homogeneous holomorphic vector bundle.

As $[\mu]$ is unitary and K permutes $\{gS_\Phi\}$ transitively, $\mathcal{V}_{\mu,\sigma}$ carries a K-invariant hermitian metric, which we use without further comment.

Every $y \in Y_\Phi$ has expression $y = gx_\Phi$, and that gives a subspace of the complexified tangent space of Y_Φ by

(1.2.14a) T_{gx_Φ} : holomorphic tangent space to gS_Φ at gx_Φ.

These spaces give a sub-bundle of the complexified tangent bundle to Y_Φ, which we denote

(1.2.14b) $\mathcal{T} \to Y_\Phi$ G-homogeneous, holomorphic over each gS_Φ.

That in turn gives us the space

(1.2.14c) $A^{0,q}(\mathcal{V}_{\mu,\sigma}) = \{C^\infty \text{ sections of } \mathcal{V}_{\mu,\sigma} \otimes \Lambda^q \bar{\mathcal{T}}^* \to Y_\Phi\}$,

whose elements we call the C^∞ <u>partially</u>-(0,q)-<u>forms</u> on Y_Φ with values in $\mathcal{V}_{\mu,\sigma}$. We have a K-invariant hermitian metric on $\mathcal{T}$, thus also on $\mathcal{V}_{\mu,\sigma} \otimes \Lambda^q \bar{\mathcal{T}}^*$, and that specifies Kodaira-Hodge operators

(1.2.15a) $A^{0,q}(\mathcal{V}_{\mu,\sigma}) \xrightarrow{\#} A^{n,n-q}(\mathcal{V}^*_{\mu,\sigma}) \xrightarrow{\tilde{\#}} A^{0,q}(\mathcal{V}_{\mu,\sigma})$

where $n = \dim_{\mathbb{C}} S_\Phi$. Similarly the $\bar{\partial}$ operator of X_Φ induces the $\bar{\partial}$-operators on each of the gS_Φ and they give

(1.2.15b) $A^{0,q}(\mathcal{V}_{\mu,\sigma}) \xrightarrow{\bar{\partial}} A^{0,q+1}(\mathcal{V}_{\mu,\sigma})$.

Now $A^{0,q}(\mathcal{V}_{\mu,\sigma})$ is a pre-Hilbert space with inner product

(1.2.16c) $\langle \alpha,\beta \rangle = \int_{K/M} d(kM) \int_{kS_\Phi} \alpha \wedge \# \beta$

where $\wedge$ indicates exterior product followed by contraction $V_\mu \otimes V^*_\mu \to \mathbb{C}$. Denote

(1.2.17a) $L_2^{0,q}(\mathcal{V}_{\mu,\sigma})$: Hilbert space completion of $A^{0,q}(\mathcal{V}_{\mu,\sigma})$.

Then $\bar{\partial}$ has formal adjoint there, given by $\bar{\partial}^* = -\tilde{\#}\bar{\partial}\#$, and we have an essentially self adjoint "partial Kodaira-Hodge-Laplace operator"

(1.2.17b) $\square = (\bar{\partial} + \bar{\partial}^*)^2 = \bar{\partial}\bar{\partial}^* + \bar{\partial}^*\bar{\partial}$ on $L_2^{0,q}(\mathcal{V}_{\mu,\sigma})$.

Our <u>partially</u>-<u>harmonic</u>-(0,q)-<u>forms</u> on Y_Φ with values in $\mathcal{V}_{\mu,\sigma}$ are the elements of

(1.2.17c) $H^{0,q}(\mathcal{V}_{\mu,\sigma}) = \{\omega \in L_2^{0,q}(\mathcal{V}_{\mu,\sigma}) : \Box\omega = 0\}$.

Observe that the partially-harmonic space (1.2.17c) consists of all Borel-measurable sections ω of $\mathcal{V}_{\mu,\sigma} \otimes \Lambda^q \bar{\mathcal{T}}^* \to Y_\Phi$ (i.e. measurable partially-(0,q)-forms) such that (i) for almost every $k \in K$, $\omega|_{kS_\Phi}$ is harmonic in the ordinary sense, and (ii) the L_2-norms of the $\omega|_{kS_\Phi}$ satisfy $\int_{K/M} \|\omega|_{kS}\|^2 d(kM) < \infty$. In particular,

(1.2.18a) $H^{0,q}(\mathcal{V}_{\mu,\sigma})$ is a closed subspace of $L_2^{0,q}(\mathcal{V}_{\mu,\sigma})$ and

(1.2.18b) G acts on $H^{0,q}(\mathcal{V}_{\mu,\sigma})$ by a continuous representation $\pi^q_{\mu,\sigma}$.

Now we have our geometric realizations:

1.2.19. Theorem. *The representation* $\pi^q_{\mu,\sigma}$ *of* G *on* $H^{0,q}(\mathcal{V}_{\mu,\sigma})$ *is unitary. If* $[\mu] = [\chi \otimes \mu^0]$ *as in* (1.1.11), *and if* β *is the highest weight of* μ^0, *then* $[\pi^q_{\mu,\sigma}]$ *is specified as follows.*

1. *If* $\langle \beta+\rho_{\mathfrak{t}}, \phi\rangle = 0$ *for some* $\phi \in \Sigma^+_{\mathfrak{t}}$, *then* $H^{0,q}(\mathcal{V}_{\mu,\sigma}) = 0$ *for all integers* q.

2. *If* $\langle \beta+\rho_{\mathfrak{t}}, \phi\rangle \neq 0$ *for every* $\phi \in \Sigma^+_{\mathfrak{t}}$, *define* q_0 *to be the number of roots* $\phi \in \Sigma^+_{\mathfrak{t}}$ *for which* $\langle \beta+\rho_{\mathfrak{t}}, \phi\rangle$ *is negative, and let* ν *be the unique element of* $L^+_{\mathfrak{m}}$ *that is conjugate to* $\beta+\rho_{\mathfrak{t}}$ *by an element of the Weyl group of* (M^0, T^0). *Then*

$$[\pi^{q_0}_{\mu,\sigma}] = [\pi_{\chi,\nu,\sigma}] \text{ \textit{principal series class},}$$

and $H^{0,q}(\mathcal{V}_{\mu,\sigma}) = 0$ *for every integer* $q \neq q_0$.

In particular, given a principal series class $[\pi_{\chi,\nu,\sigma}]$, *we can realize it on* $H^{0,0}(\mathcal{V}_{\mu,\sigma})$ *where* $\mu = [\chi \otimes \mu^0]$ *and* μ^0 *has highest weight* $\nu - \rho_{\mathfrak{t}}$.

Proof. Let $\tilde{\pi}^q_{\mu,\sigma}$ denote the representation of G on $L_2^{0,q}(\mathcal{V}_{\mu,\sigma})$. Factor $\gamma_{\mu,\sigma} = {}'\gamma_{\mu,\sigma} \cdot e^{\rho_{\mathfrak{a}}}$ where ${}'\gamma_{\mu,\sigma}(uan) = e^{i\sigma}(a)\mu(u)$ unitary, and

$e^{\rho_{\mathfrak{a}}}(uan) = e^{\rho_{\mathfrak{a}}}(a)$ inverse of the square root of the action of uan of the volume element of $K/U_\Phi = Y_\Phi$. Thus $\tilde{\pi}^q_{\mu,\sigma} = \mathrm{Ind}_{U_\Phi AN\uparrow G}({}'\gamma_{\mu,\sigma})$ unitary. In particular its subrepresentation $\pi^q_{\mu,\sigma}$ is unitary.

Abbreviate $V^q_\mu = V_\mu \otimes \Lambda^q(\bar{T}^*_{x_\Phi})$. That identifies $L_2^{0,q}(\mathcal{V}_{\mu,\sigma})$ with the space of all measurable $f\colon G \to V^q_\mu$ such that

$$f(guan) = (\gamma_{\mu,\sigma}(uan)^{-1} \otimes \Lambda^q ad(uan)\big|_{\mathfrak{n}}^{-1})f(g) \quad \text{and} \quad \int_{K/U_\Phi} \|f(k)\|^2 d(kU_\Phi) < \infty.$$

Decompose

$$\tilde{\pi}^q_{\mu,\sigma} = \mathrm{Ind}_{B\uparrow G}(\tilde{\psi}) \quad \text{where} \quad \tilde{\psi} = \mathrm{Ind}_{U_\Phi AN\uparrow B}({}'\gamma_{\mu,\sigma}).$$

Then the representation space elements annihilated by the partial Laplacian $\square$, correspond to the subspace of the representation space of $\tilde{\psi}$ annihilated by the full Laplacian of $\mathcal{V}_{\mu,\sigma}|_{S_\Phi}$. Thus

$$\pi^q_{\mu,\sigma} = \mathrm{Ind}_{B\uparrow G}(\psi) \quad \text{where } \psi \text{ represents } B \text{ on } H^{0,q}(\mathcal{V}_{\mu,\sigma}|_{S_\Phi}).$$

Let us denote

$$\eta^q_\mu\colon \text{ representation of } M \text{ on } H^{0,q}(\mathcal{V}_\mu).$$

Then $\psi(man) = \eta^q_\mu(m)e^{i\sigma}(a)$. The assertion now follows from definition (1.2.8) and from Proposition 1.1.12. *q.e.d.*

§2. General Notion of Relative Discrete Series

We write down the basic facts concerning square integrable representations of locally compact unimodular groups.

We use the following notation concerning a locally compact group G. Left Haar measure is denoted dg, so the left regular representation of G on $L_2(G)$ is

$$[\ell(g)f](x) = f(g^{-1}x).$$

If π is an irreducible unitary representation of G then H_π denotes its representation space, $[\pi]$ denotes its unitary equivalence class, and the coefficients of $[\pi]$ are the functions

$$\phi_{\xi,\eta}(g) = \langle \xi, \pi(g)\eta \rangle \quad \text{where} \quad \xi, \eta \in H_\pi$$

on G. The set of all equivalence classes of irreducible unitary representations of G is denoted $\hat{G}$.

2.1. Let G be a unimodular locally compact group and Z a closed normal abelian subgroup. Then Z has left regular representation $\ell^Z = \int_{\hat{Z}} \zeta d\zeta$. Thus the left regular representation of G is

$$\ell = \mathrm{Ind}_{\{1\}\uparrow G}(1) = \mathrm{Ind}_{Z\uparrow G}(\ell^Z) = \int_{\hat{Z}} \mathrm{Ind}_{Z\uparrow G}(\zeta)d\zeta.$$

If $\zeta \in \hat{Z}$ we define $\ell_\zeta = \mathrm{Ind}_{Z\uparrow G}(\zeta)$. It is the left regular representation of G on the Hilbert space

(2.1.1) $L_2(G/Z,\zeta) = \{f\colon G \to C\colon f(gz) = \zeta(z)^{-1}f(g) \text{ and } \int_{G/Z} |f(g)|^2 d(gZ) < \infty\}$.

With that notation,

(2.1.2) $$\ell = \int_{\hat{Z}} \ell_\zeta d\zeta \quad \text{and} \quad L_2(G) = \int_{\hat{Z}} L_2(G/Z,\zeta)d\zeta.$$

Given $\zeta \in \hat{Z}$ we define

(2.1.3) $$\hat{G}_\zeta = \{[\pi] \in \hat{G}\colon \ \zeta \text{ is a subrepresentation of } \pi|_Z\}.$$

Thus ℓ_ζ is a direct integral over $\hat{G}_\zeta$. We say that a class $[\pi] \in \hat{G}$ is ζ-discrete if π is a subrepresentation of ℓ_ζ. The ζ-discrete classes form the ζ-discrete series $\hat{G}_{\zeta\text{-disc}}$. Note $\hat{G}_{\zeta\text{-disc}} \subset \hat{G}_\zeta$. Finally the relative (to Z) discrete series is

(2.1.4) $$\hat{G}_{disc} = \bigcup_{\zeta\in\hat{Z}} \hat{G}_{\zeta\text{-disc}}.$$

2.2. Suppose that the closed normal abelian subgroup Z is central in G. Then every class $[\pi] \in \hat{G}$ specifies a character $\zeta_\pi \in \hat{Z}$ by: $\pi|_Z$ is a multiple of ζ_π. In other words $\hat{G} = \bigcup_{\hat{Z}} \hat{G}_\zeta$ is disjoint.

Let $[\pi] \in \hat{G}_\zeta$. Since Z is central, any coefficient $\phi = \phi_{\xi,\eta}$ of $[\pi]$ satisfies $\phi(gz) = \langle \xi,\pi(g)\pi(z)\eta\rangle = \overline{\zeta(z)}\langle \xi,\pi(g)\eta\rangle = \zeta(z)^{-1}\phi(g)$. If $|\phi| \in L_2(G/Z)$ then we conclude $\phi \in L_2(G/Z,\zeta)$.

Here are the basic facts on the ζ-discrete series (Z central) of G.

If $[\pi] \in \hat{G}$ and $\zeta \in \hat{Z}$ then the following conditions are equivalent.

(2.2.1a) There exists $0 \neq \xi \in H_\pi$ with $\phi_{\xi,\xi} \in L_2(G/Z,\zeta)$.

(2.2.1b) If $\xi, \eta \in H_\pi$ then $\phi_{\xi,\eta} \in L_2(G/Z,\zeta)$.

(2.2.1c) $[\pi]$ is the ζ-discrete series $\hat{G}_{\zeta\text{-disc}}$.

Under conditions (2.2.1) there is a positive real number $\deg(\pi)$, called the formal degree of $[\pi]$, such that

(2.2.2a) $\langle \phi_{\xi,\eta},\phi_{\alpha,\beta}\rangle = \deg(\pi)^{-1}\langle \xi,\alpha\rangle\overline{\langle \eta,\beta\rangle}$ for $\xi, \eta, \alpha, \beta \in H_\pi$.

Further if $[\pi'] \neq [\pi]$, both in $\hat{G}_{\zeta\text{-disc}}$, then

(2.2.2b) $\langle \phi_{\xi,\eta},\phi_{\xi',\eta'}\rangle = 0$ for $\xi, \eta \in H_\pi$ and $\xi', \eta' \in H_{\pi'}$.

If G is compact, $Z = \{1\}$ and we normalize $\int_G dg = 1$, then $\hat{G} = \hat{G}_{disc}$, $\deg(\pi)$ is the degree in the usual sense, and (2.2.2) reduces to the Frobenius-Schur Relations. See §2.4 below.

One defines $L_p(G/Z,\zeta)$, as in (2.1.1), by integration over G/Z. Since

Z is central,

$$(f*h)(x) = \int_{G/Z} f(g)h(g^{-1}x)d(gZ)$$

gives a well defined convolution product

$$L_1(G/Z,\zeta) \times L_p(G/Z,\zeta) \to L_p(G/Z,\zeta).$$

It is convenient to express (2.2.2a) in the form

(2.2.2c) $$\phi_{\xi,\eta} * \phi_{\alpha,\beta} = \deg(\pi)^{-1}\langle\xi,\beta\rangle\phi_{\alpha,\eta} \in L_2(G/Z,\zeta).$$

The results (2.2.1) and (2.2.2) are due to Godement [8] for compact Z, to Harish-Chandra [19] for semisimple G. One can get them, either by imitating Dixmier's exposition [6,§14] of Godement, or by applying Rieffel's results [41] to the convolution algebra $L_1(G/Z,\zeta) \cap L_2(G/Z,\zeta)$.

2.3. We relax the hypothesis on the closed normal abelian subgroup Z, to

(2.3.1) G has a finite index subgroup J that centralizes Z.

Without loss of generality we may assume $Z \subset J$ and J normal in G.

If $\zeta \in \hat{Z}$ we denote $\ell_\zeta^J = \mathrm{Ind}_{Z\uparrow J}(\zeta)$. Thus $\ell_\zeta = \mathrm{Ind}_{J\uparrow G}(\ell_\zeta^J)$. Since $|G/J| < \infty$, the restriction

$$\ell_\zeta|_J = \sum_{xJ\in G/J} \ell_{\zeta_x}^J \quad \text{finite sum,} \quad \zeta_x(z) = \zeta(x^{-1}zx).$$

These facts tell us

(2.3.2a) $\hat{G}_{disc} = \{[\pi]\} \in \hat{G}\colon \pi|_J$ has a subrepresentation $[\psi] \in \hat{J}_{disc}\}$

and

(2.3.2b) $\hat{J}_{disc} = \{[\psi] \in \hat{J}\colon \psi$ is a subrepresentation of $\pi|_J$ for some $[\pi] \in \hat{G}_{disc}\}$.

In our applications, if $[\psi] \in \hat{J}_{disc}$ then $\mathrm{Ind}_{J\uparrow G}(\psi)|_J$ is a sum of mutually inequivalent representations, so $[\psi] \to [\mathrm{Ind}_{J\uparrow G}(\psi)]$ will map $\hat{J}_{disc}$ onto $\hat{G}_{disc}$.

2.4. We eventually realize the relative discrete series $\hat{G}_{disc}$ on homogeneous vector bundles over coset spaces G/U where $Z \subset U$ and U/Z is compact. Thus we need a mild extension of the Peter-Weyl theorem. The extension is sufficiently pedestrian that we just derive it rather than extract it from the more elaborate Grosser-Moskowitz theory ([10], [11], [12]).

Let U be a locally compact group and Z a closed central subgroup such that U/Z is compact.

2.4.1. Lemma. *Every class* $[\chi] \in \hat{U}$ *is finite dimensional.*

Proof. Let $S = \{s \in C: |s| = 1\}$ circle group and $1_S \in \hat{S}$ the character $1_S(s) = s$. If $\zeta \in \hat{Z}$ we define(1)

$$U[\zeta] = \{S \times U\}/\{(\zeta(z)^{-1}, z): z \in Z\}.$$

If $[\chi] \in \hat{U}_\zeta$ now $[1_S \otimes \chi] \in (S \times U)^\wedge$ factors through $U[\zeta]$. But $U[\zeta]$ is an extension

$$1 \to S \to U[\zeta] \to U/Z \to 1$$

of a compact group by a compact group, hence is compact. Thus $1_S \otimes \chi$ is finite dimensional. *q.e.d.*

2.4.2. Lemma. *If* $\zeta \in \hat{Z}$ *then* $\hat{U}_\zeta = \hat{U}_{\zeta\text{-disc}}$.

Proof. If ϕ is a coefficient of $[\chi] \in \hat{U}_\zeta$ then $|\phi|$ is a continuous function on the compact space U/Z. Thus $\phi \in L_2(U/Z,\zeta)$ and so $[\chi] \in \hat{U}_{\zeta\text{-disc}}$. *q.e.d.*

2.4.3. Lemma. *Fix* $\zeta \in \hat{Z}$. *If* $[\chi] \in \hat{U}_\zeta$ *let* V_χ *denote its representation space and* E_χ *its space of coefficients. Then*

$$L_2(U/Z,\zeta) = \sum_{\hat{U}_\zeta} E_\chi = \sum_{\hat{U}_\zeta} V_\chi \otimes V_\chi^*$$

orthogonal direct sum, where the left action of U *on* $E_\chi = V_\chi \otimes V_\chi^*$ *is*

---

(1) This construction is important for us. See §§3.3 and 3.4 below.

$\chi \otimes 1$ *and the right action is* $1 \otimes \chi^*$.

Proof. Let $q: U \to U[\zeta]$ be restriction of the projection $S \times U \to U[\zeta]$ of Lemma 2.4.1. Then $f \to f \cdot q$ is an equivariant isometry of $L_2(U[\zeta]/S, 1_S)$ onto $L_2(U/Z, \zeta)$. The Peter-Weyl Theorem for $U[\zeta]$, insofar as it concerns $\widehat{U[\zeta]}_{1_S}$, now carries directly over to our assertions. *q.e.d.*

The space of compactly supported continuous functions on U is denoted $C_c(U)$. If $f \in C_c(U)$ then $\chi(f): V_\chi \to V_\chi$ by $\chi(f)v = \int_U f(u)\chi(u)v\, du$. Also, suppose Haar measures normalized by

$$\int_U du = \int_Z dz \int_{U/Z} d(uZ) \quad \text{and} \quad \int_{U/Z} 1 d(uZ) = 1.$$

2.2.4. Proposition. *Let* $\zeta \in \hat{Z}$ *and* $[\chi] \in \hat{U}_\zeta$. *Then*

$$\text{trace } \chi(f) = \int_U f(u) \cdot \text{trace } \chi(u) du \quad \textit{for} \quad f \in C_c(U)$$

and

$$\dim \chi (= \dim V_\chi) \quad \textit{is the formal degree} \quad \deg \chi.$$

Finally, orthogonal projection of $L_2(U/Z, \zeta)$ *to* E_χ *is right or left convolution by* $(\dim \chi)$ trace $\bar{\chi}$.

Proof. Let $\{v_1, \ldots, v_n\}$ be an orthonormal basis of V_χ and $\phi_{ij}(u) = \langle v_i, \chi(u)v_j \rangle$. For the first assertion,

$$\text{trace } \chi(f) = \sum \langle \chi(f)v_i, v_i \rangle = \sum \int_U \langle f(u)\chi(u)v_i, v_i \rangle du$$

$$= \int_U f(u)\left(\sum \langle \chi(u)v_i, v_i \rangle\right) du = \int_U f(u) \text{trace } \chi(u) du.$$

For the second assertion use the trick $q: U \to U[\zeta]$ of Lemma 2.4.3: then $\dim \chi = \dim(1_S \otimes \chi) = \deg(1_S \otimes \chi) = \deg(\chi)$. Or one can use (2.2.2a) to calculate

$$1 = \langle \text{trace } \chi, \text{trace } \chi \rangle = \sum_{i,j=1}^{n} \langle \phi_{ii}, \phi_{jj} \rangle = n/\deg(\chi).$$

The last assertion now follows from $q: U \to U[\zeta]$, and also follows from (2.2.2c). *q.e.d.*

Combining (2.1.2), Lemma 2.4.3 and Proposition 2.4.4 we have the Plancherel formula for U:

2.4.5. Proposition: $L_2(U) = \int_{\hat{Z}} \left\{ \sum_{\hat{U}_\zeta} V_\chi \otimes V_\chi^* \right\} d\zeta.$

More precisely, if $f \in C_c(U)$, *and if for* $\zeta \in \hat{Z}$ *we define* $f_\zeta(u) = \int_Z f(uz)\zeta(z)dz$, *then*

(2.4.6) $$f(1) = \int_{\hat{Z}} \left\{ \sum_{\chi \in \hat{U}_\zeta} \operatorname{trace} \chi(f_\zeta) \dim \chi \right\} d\zeta.$$

2.5. We use the method of §2.4 to prove an exact version of Frobenius' Reciprocity Theorem which we need in [57].

Let G be a locally compact unimodular group, Z a closed central subgroup, and U/Z a compact subgroup of G/Z. From §2.4 we know that $\hat{U} = \hat{U}_{disc}$ and that every class $[\chi] \in \hat{U}$ is finite dimensional. The case $Z = \{1\}$ of the following theorem is due to Kunze [55] and Wawrzyńczyk [56]; our proof is a reduction to their case. Frobenius' original theorem is the case where G is finite.

2.5.1. Theorem. *Let* $[\pi] \in \hat{G}_{disc}$ *and* $[\chi] \in \hat{U}$. *Then multiplicities satisfy*

$$m(\chi, \pi|_U) = m(\pi, \operatorname{Ind}_{U\uparrow G}(\chi)).$$

Further, these multiplicities vanish unless $[\pi] \in \hat{G}_\zeta$ *and* $[\chi] \in \hat{U}_\zeta$ *for the same character* $\zeta \in \hat{Z}$.

Proof. Define $\zeta, \zeta' \in \hat{Z}$ by $[\chi] \in \hat{U}_\zeta$ and $[\pi] \in \hat{G}_{\zeta'}$. Then $(\pi|_U)|_Z$ is a multiple of ζ' and $\operatorname{Ind}_{U\uparrow G}(\chi)|_Z$ is a multiple of ζ. Thus both multiplicities vanish unless $\zeta' = \zeta$.

Now suppose $\zeta' = \zeta$. As in the proof of Lemma 2.4.1 let $p: G \to G[\zeta]$

denote the composition

$$G \to S \times G \to (S \times G)/\{(\zeta(z)^{-1}, z) : z \in Z\} = G[\zeta]$$

and let $q = p|_U : U \to U[\zeta]$ denote the composition

$$U \to S \times U \to (S \times U)/\{(\zeta(z)^{-1}, z) : z \in Z\} = U[\zeta].$$

Then we have bijective correspondences (see §3.3 below)

$$\varepsilon_G : G[\zeta]_1^{\wedge} \to \hat{G}_\zeta \quad \text{by} \quad \varepsilon_G[\psi] = [\psi \cdot p]$$

and

$$\varepsilon_U : U[\zeta]_1^{\wedge} \to \hat{U}_\zeta \quad \text{by} \quad \varepsilon_U[\phi] = [\phi \cdot q].$$

Use these correspondences to express

(2.5.2) $[\pi] = \varepsilon_G[\psi] = [\psi \cdot p]$ and $[\chi] = \varepsilon_U[\phi] = [\phi \cdot q]$.

Note $H_\pi = H_\psi$. If $\xi, \eta \in H_\pi$ and Haar measures are properly normalized, then $G/Z = G[\zeta]/S$ gives us

$$\int_{G[\zeta]} |\langle \xi, \psi(x)\eta \rangle|^2 \, dx = \int_{G[\zeta]/S} |\langle \xi, \psi(x)\eta \rangle|^2 \, d(xS)$$

$$= \int_{G/Z} |\langle \xi, \pi(g)\eta \rangle|^2 \, d(gZ) < \infty.$$

Thus $[\psi]$ is in the ordinary discrete series of $G[\zeta]$. Also $U[\zeta]$ is compact because $U[\zeta]/S = U/Z$. Now R. A. Kunze's result [55,Theorem 3] and A. Wawrzyńczyk's result [56,Theorem 2.1] each say that

(2.5.3) $m(\phi, \psi|_{U[\zeta]}) = m(\psi, \mathrm{Ind}_{U[\zeta]\uparrow G[\zeta]}(\phi))$.

From (2.5.2), $(\psi|_{U[\zeta]}) \cdot q = (\psi \cdot p)|_U = \pi|_U$. Now

(2.5.4) $m(\chi, \pi|_U) = m(\phi \cdot q, (\psi|_{U[\zeta]}) \cdot q) = m(\phi, \psi|_{U[\zeta]})$.

Similarly $\{\mathrm{Ind}_{U[\zeta]\uparrow G[\zeta]}(\phi)\} \cdot q = \mathrm{Ind}_{U\uparrow G}(\phi \cdot p) = \mathrm{Ind}_{U\uparrow G}(\chi)$, so

(2.5.5) $m(\psi, \mathrm{Ind}_{U[\zeta]\uparrow G[\zeta]}(\phi)) = m(\pi, \mathrm{Ind}_{U\uparrow G}(\chi))$.

Theorem 2.5.1 follows by combining (2.5.3), (2.5.4) and (2.5.5). *q.e.d.*

§3. Relative Discrete Series for Reductive Groups.

We extend Harish-Chandra's theory [24] of discrete series representations to a class of reductive groups that contains all connected reductive groups and has certain hereditary properties. That class of groups is described in §3.1. Some of the main points of Harish-Chandra's general character theory are recalled in §3.2. In §3.3 we work out a method for reducing representation-theoretic questions on connected reductive groups to the case of compact center. We use the method in §3.4 to extend Harish-Chandra's discrete series theory to all connected reductive Lie groups, in particular to all connected semisimple groups. Finally in §3.5, specifically in Theorems 3.5.8 and 3.5.9, we obtain the discrete series for the class described in §3.1.

A number of technical lemmas are done in greater generality than is immediately necessary, so that they do not have to be repeated in §§4 and 5.

3.1. From now on, G is a reductive Lie group in the sense that its Lie algebra

$$\mathfrak{g} = \mathfrak{c} \oplus \mathfrak{g}_1 \quad \text{with } \mathfrak{c} \text{ central and } \mathfrak{g}_1 = [\mathfrak{g},\mathfrak{g}] \text{ semisimple.}$$

For technical convenience we assume

(3.1.1) if $g \in G$ then $\mathrm{ad}(g)$ is an inner automorphism of $\mathfrak{g}_C$.

Finally we suppose that G has a closed normal abelian subgroup Z such that

(3.1.2a) Z centralizes the identity component G^0 of G,

(3.1.2b) ZG^0 has finite index in G, and

(3.1.2c) $Z \cap G^0$ is co-compact in the center Z_{G^0} of G^0.

If $|G/G^0| < \infty$ then Z_{G^0} satisfies (3.1.2).

It is straightforward to see that the discrete series of G relative to

Z is independent of choice of group Z which satisfies (3.1.2).

We will use the notation

(3.1.3a) $G^\dagger = \{g \in G\colon \mathrm{ad}(g)$ is an inner automorphism of $G^0\}$.

Then, of course,

(3.1.3b) $G^\dagger = Z_G(G^0)G^0$ where $Z_G(G^0) = \{g \in G\colon g$ centralizes $G^0\}$.

Also observe that

(3.1.3c) $Z_G(G^0)/Z$ is compact and $G^\dagger/ZG^0$ is finite.

3.2. G/Z has only finitely many topological components by (3.1.2b), so we can choose

(3.2.1) K/Z: maximal compact subgroup of G/Z.

From general theory, $K^0 = K \cap G^0$ and K meets every topological component of G. Also $Z_{G^0} \subset K$ by (3.1.2c).

The following fact is the basis of Harish-Chandra's global character theory. There is an integer $n_G \geqslant 1$ such that, if $[\kappa] \in \hat{K}$ and $[\pi] \in \hat{G}$ then

(3.2.2) the multiplicity $m(\kappa,\pi|_K) \leqslant n_G \dim(\kappa) < \infty$.

This is proved by Harish-Chandra [15] for connected reductive groups, so we know it for G^0. Then if $[\kappa_1] \in (ZK^0)^\wedge$ and $[\pi_1] \in (ZG^0)^\wedge$ we have $\zeta, \zeta' \in \hat{Z}$, $[\kappa_0] \in \hat{K}^0$ and $[\pi_0] \in \hat{G}^0$ such that $\kappa_1 = \zeta \otimes \kappa_0$ and $\pi_1 = \zeta \otimes \pi_0$. Now, writing n for n_{G^0},

$$m(\kappa_1,\pi_1|_{ZK^0}) \leqslant m(\kappa_0,\pi_0|_{K^0}) \leqslant n\cdot\dim(\kappa_0) = n\cdot\dim(\kappa_1)$$

so we know (3.2.2) for ZG^0. Finally, if $[\kappa] \in \hat{K}$ and $[\pi] \in \hat{G}$ we decompose

$$\kappa|_{ZK^0} = \sum \kappa_i \quad\text{and}\quad \pi|_{ZG^0} = \sum \pi_j$$

with κ_i and π_j irreducible. Then

$$m(\kappa,\pi|_K) \leq \sum_{i,j} m(\kappa_i,\pi_j|_{ZK^0}) \leq \sum_{i,j} n\cdot\dim(\kappa_i)$$
$$= \sum_i n\cdot\dim(\kappa) \leq (n\cdot|G/ZG^0|)\dim(\kappa)$$

so (3.2.2) holds in our general case.

The first consequence of (3.2.2) is that the group G is CCR. That is, if $[\pi] \in \hat{G}$ and $f \in L_1(G)$ then $\pi(f) = \int_G f(g)\pi(g)dg$ is a compact operator on H_π. In particular G is type I.

The second consequence of (3.2.2) is that $\pi(f)$ is of trace class for $f \in C_c^\infty(G)$, and that

(3.2.3) Θ_π: $C_c^\infty(G) \to C$ by $\Theta_\pi(f) = \text{trace } \pi(f)$

is a Schwartz distribution on G. Θ_π is the global character or distribution character of $[\pi]$. Classes $[\pi] = [\pi']$ if and only if $\Theta_\pi = \Theta_{\pi'}$.

A differential operator z on G has transpose given by

$\int_G [{}^t z(f)](g)h(g)dg = \int_G f(g)[z(h)](g)dg$ for $f, h \in C_c^\infty(G)$. It acts on distributions by $(z\Theta)(f) = \Theta({}^t z\cdot f)$ for $f \in C_c^\infty(G)$. Given Θ now $z \to z\Theta$ is linear in z. A distribution Θ on G is invariant if

$\Theta(f) = \Theta(f\cdot \mathrm{ad}(g))$ for all $f \in C_c^\infty(G)$ and $g \in G$.

The universal enveloping algebra $\mathfrak{G}$ of $\mathfrak{g}_C$ is the associative algebra of all left-invariant differential operators on G. The center $\mathfrak{Z}$ of $\mathfrak{G}$ consists of the bi-invariant operators; this is equivalent to (3.1.1). A distribution Θ on G is an eigendistribution of $\mathfrak{Z}$ if $\dim \mathfrak{Z}(\Theta) \leq 1$. In that case, using commutativity of $\mathfrak{Z}$, we have an algebra homomorphism χ_Θ: $\mathfrak{Z} \to C$ defined by $z\Theta = \chi_\Theta(z)\Theta$.

Let $[\pi] \in \hat{G}$. Its distribution character Θ_π is an invariant eigendistribution. The associated homomorphism

(3.2.4) $\chi_\pi: \mathfrak{Z} \to C$ by $z\Theta_\pi = \chi_\pi(z)\Theta_\pi$

is the <u>infinitesimal</u> <u>character</u> of $[\pi]$.

Choose $\mathfrak{h} \subset \mathfrak{g}$ Cartan subalgebra. Let $\underline{I}(\mathfrak{h}_C)$ denote the algebra of all polynomials on $\mathfrak{h}_C^*$ that are invariant under the Weyl group $W(\mathfrak{g}_C, \mathfrak{h}_C)$. Harish-Chandra [20] found an isomorphism $\gamma: \mathfrak{Z} \to \underline{I}(\mathfrak{h}_C)$ with these properties. If $\gamma \in \mathfrak{h}_C^*$ then

(3.2.5) $\chi_\lambda: \mathfrak{Z} \to C$ by $\chi_\lambda(z) = [\gamma(z)](\lambda)$

is a homomorphism. Every homomorphism $\mathfrak{Z} \to C$ is one of the χ_λ. And $\chi_\lambda = \chi_{\lambda'}$ precisely when λ and λ' are $W(\mathfrak{g}_C, \mathfrak{h}_C)$-equivalent.

Harish-Chandra [20] used the differential equations $z\Theta_\pi = \chi_\pi(z)\Theta_\pi$ of (3.2.4) and the description (3.2.5) of χ_π, to show that Θ_π is a locally L_1 function on G which is analytic on the <u>regular</u> <u>set</u> G'. Here G' consists of all $g \in G$ whose fixed point set on $\mathfrak{g}$ is a Cartan subalgebra of $\mathfrak{g}$. G' is dense and open in G, and its complement has measure zero.

The differential equations also show that at most finitely many classes in $\hat{G}$ can have the same infinitesimal character.

<u>3.3</u>. We now describe a method for reducing questions of harmonic analysis on the groups ZG^0 and G^0, to the same questions for connected reductive Lie groups with compact center. Among other things, this will allow us to extend some of Harish-Chandra's results [24] on discrete series to the group G^0 in §3.4, and then to G in §3.5.

As in §2.4, $S = \{s \in C: |s| = 1\}$ circle group and $1 = 1_S \in \hat{S}$ is given by $1(s) = s$. If $\zeta \in \hat{Z}$ we have the quotient group

(3.3.1a) $G[\zeta] = \{S \times ZG^0\}/\{(\zeta(z)^{-1}, z): z \in Z\}$.

Although we do not need the fact, we remark that $G[\zeta]$ is the Mackey central extension

(3.3.1b) $1 \to S \to G[\zeta] \to ZG^0/Z \to 1$

for the coboundary of ζ as normalized multiplier on ZG^0/Z. At any event, $G[\zeta]$ is a connected reductive Lie group with Lie algebra $\mathfrak{s} \oplus (\mathfrak{g}/\mathfrak{z})$ and with compact center.

3.3.2. Lemma. *Let* $p\colon ZG^0 \to G[\zeta]$ *denote the restriction of the projection* $S \times ZG^0 \to G[\zeta]$. *Then*(2) $f \to f\cdot p$ *gives an equivariant isometry of* $L_2(G[\zeta]/S,1)$ *onto* $L_2(ZG^0/Z,\zeta)$.

Proof. View $f \in L_2(G[\zeta]/S,1)$ as a function on $S \times ZG^0$. Then, if $g \in ZG^0$ and $z \in Z$,

$$(f\cdot p)(gz) = f(1,gz) = f(\zeta(z),g) = \zeta(z)^{-1}f(1,g) = \zeta(z)^{-1}[f\cdot p](g)$$

and

$$\int_{ZG^0/Z} |(f\cdot p)(g)|^2 d(gZ) = \int_{(S\times ZG^0)/S\times Z} |f(s,g)|^2 d(sS\times gZ)$$

$$= \int_{G[\zeta]/S} |f(\bar{g})|^2 d(\bar{g}S).$$

Thus $f \to f\cdot p$ is an isometric injection of $L_2(G[\zeta]/S,1)$ into $L_2(ZG^0/Z,\zeta)$. But it is surjective because any $f' \in L_2(ZG^0/Z,\zeta)$ is of the form $f\cdot p$ where f is defined on $S \times ZG^0$ by $f(s,g) = s^{-1}f'(g)$. *q.e.d.*

3.3.3. Theorem. *There is a well-defined bijection* $\varepsilon = \varepsilon_\zeta\colon G[\zeta]^\wedge_1 \to (ZG^0)^\wedge_\zeta$ *given by* $\varepsilon[\psi] = [\psi\cdot p]$. *It maps* $G[\zeta]^\wedge_{1\text{-disc}}$ *onto* $(ZG^0)^\wedge_{\zeta\text{-disc}}$ *and carries Plancherel measure of* $G[\zeta]^\wedge_1$ *to Plancherel measure of* $(ZG^0)^\wedge_\zeta$. *Distribution characters satisfy* $\Theta_{\varepsilon[\psi]} = \Theta_{[\psi]}\cdot p$.

Proof. Let $[\psi] \in G[\zeta]^\wedge_1$ and view it as a representation of $S \times ZG^0$. If $z \in Z$ and $g \in ZG^0$ then

(2) Of course we assume consistent normalizations of Haar measures.

$$(\psi\cdot p)(gz) = \psi(1,gz) = \psi(\zeta(z),g) = \zeta(z)\psi(1,g)$$
$$= \zeta(z)[\psi\cdot p](g).$$

This shows that $\psi\cdot p$ has central character whose Z-restriction is ζ. It also shows, since ψ is irreducible and unitary, that $\psi\cdot p$ is irreducible and unitary. Thus $[\psi\cdot p] \in (ZG^0)^{\wedge}_{\zeta}$.

Let $[\psi], [\psi'] \in G[\zeta]^{\wedge}_1$ and $b: H_\psi \to H_{\psi'}$ isometry. If $\psi' = b\cdot\psi\cdot b^{-1}$ the above calculation shows $(\psi'\cdot p) = b\cdot(\psi\cdot p)\cdot b^{-1}$. If b intertwines $\psi\cdot p$ with $\psi'\cdot p$ the same calculation shows that it intertwines $\psi|_{ZG^0}$ with $\psi'|_{ZG^0}$. Then it intertwines ψ with ψ' because $\psi'(s,g) = s\psi'(1,g) = sb\psi(1,g)b^{-1} = b\cdot s\psi(1,g)\cdot b^{-1} = b\cdot\psi(s,g)\cdot b^{-1}$. Now $\varepsilon_\zeta: G[\zeta]^{\wedge}_1 \to (ZG^0)^{\wedge}_{\zeta}$ is a well-defined bijection.

We reduce the proof of Theorem 3.3.3 to the case where $Z \subset G^0$, i.e. where ZG^0 is connected. Let $\zeta^0 = \zeta|_{Z\cap G^0}$ and observe from (3.3.1) that the inclusion $G^0 \to G$ induces a commutative diagram

$$\begin{array}{ccccccccc} 1 & \to & S & \to & G^0[\zeta^0] & \to & G^0/Z\cap G^0 & \to & 1 \\ & & \downarrow & & \downarrow{\scriptstyle a} & & \downarrow & & \\ 1 & \to & S & \to & G[\zeta] & \to & ZG^0/Z & \to & 1 \end{array}$$

of Lie group homomorphisms. Since $S \to S$ and $G^0/Z\cap G^0 \to ZG^0/Z$ are isomorphisms, so is $a: G^0[\zeta^0] \to G[\zeta]$. That gives a commutative diagram

$$\begin{array}{ccc} G^0[\zeta^0]^{\wedge}_1 & \xrightarrow{\varepsilon^0} & (G^0)^{\wedge}_{\zeta^0} \\ {\scriptstyle a^*}\nearrow & & \searrow{\scriptstyle b} \\ G[\zeta]^{\wedge}_1 & \xrightarrow{\quad\varepsilon\quad} & (ZG^0)^{\wedge}_{\zeta} \end{array}$$

where a^* is induced from $a: G^0[\zeta^0] \cong G[\zeta]$ and $b[\pi^0] = [\zeta \otimes \pi^0]$. Everything is conserved by a^*. Plancherel measure and relative discrete series are transported by b. If $[\pi^0] \in (G^0)^{\wedge}_{\zeta^0}$, $z \in Z$ and $g \in G^0$ then

$\Theta_{b[\pi^0]}(zg) = \zeta(z)\Theta_{\pi^0}(g)$. Thus if $\varepsilon^0: G^0[\zeta^0]^\wedge_1 \to (G^0)^\wedge_{\zeta^0}$ has the properties asserted by Theorem 3.3.3, those properties will be inherited by $\varepsilon: G[\zeta]^\wedge_1 \to (ZG^0)^\wedge_\zeta$.

Now we assume $Z \subset G^0$ so $ZG^0 = G^0$. We further reduce the proof of Theorem 3.3.3 to the case where G^0 is simply connected. Let $\tau: \tilde{G} \to G^0$ be the universal cover, $\tilde{Z} = \tau^{-1}(Z)$, and $\tilde{\zeta} = \zeta\cdot\tau$ the corresponding lift of ζ. Then one has $i: \tilde{G}[\tilde{\zeta}] \cong G[\zeta]$ induced from $1_S \times \tau$, and also a commutative diagram

$$\begin{array}{ccc} \tilde{G}[\tilde{\zeta}]^\wedge_1 & \xrightarrow{\ \tilde{\varepsilon}\ } & (\tilde{G})^\wedge_{\tilde{\zeta}} \\ {\scriptstyle i^*}\nearrow & & \nwarrow{\scriptstyle j} \\ G[\zeta]^\wedge_1 & \xrightarrow{\ \varepsilon\ } & (G^0)^\wedge_\zeta \end{array}$$

where i induces i^* and $j[\pi^0] = [\pi^0\cdot\tau]$. Everything is conserved by i^*. Plancherel measure and relative discrete series are transported by j.

We check $\Theta_{j[\psi]} = \Theta_{[\psi]}\cdot\tau$ for $[\psi] \in (G^0)^\wedge_\zeta$. It suffices to test on functions $f \in C_c^\infty(\tilde{G})$ such that G^0 contains an open set U_1 admissible for the covering $\tau: \tilde{G} \to G^0$ and the support of f is contained in a topological component U of $\tau^{-1}(U_1)$. For an appropriate C^∞ partition of unity on $\tilde{G}$ breaks up every $C_c^\infty(\tilde{G})$-function as a finite sum of such f. Now f gives $f_1 \in C_c^\infty(U_1) \subset C_c^\infty(G^0)$ by the formula $f_1(\tau x) = f(x)$ for $x \in U$. Then we calculate

$$\Theta_{j[\psi]}(f) = \text{trace} \int_U f(x)(\psi\cdot\tau)(x)dx = \text{trace} \int_U f_1(\tau x)\psi(\tau x)dx$$

$$= \text{trace} \int_{U_1} f_1(x_1)\psi(x_1)dx_1 = \Theta_{[\psi]}(f_1)$$

to conclude $\Theta_{j[\psi]} = \Theta_{[\psi]}\cdot j$.

If the map $\tilde{\varepsilon}: \tilde{G}[\tilde{\zeta}]^\wedge_1 \to (\tilde{G})^\wedge_\zeta$ has the properties claimed in Theorem

3.3.3, now those properties are shared by $\varepsilon\colon G[\zeta]_1^\wedge \to (G^0)_\zeta^\wedge$.

We are reduced to the case where $Z \subset G^0$ and G^0 is simply connected. Split $G^0 = V \times G_{ss}$ where V is a vector group, G_{ss} is semisimple, and Z is co-compact in the center $V \times Z_{ss}$ of G^0. We note that $(G^0)_\zeta^\wedge$ is disjoint union of the $(G^0)_\xi^\wedge$ as ξ ranges over $(VZ)_\zeta^\wedge$. Since VZ/Z is compact, there is no distinction of relative discrete class, and Plancherel measure on $(G^0)_\zeta^\wedge$ is just the sum of the Plancherel measures on the $(G^0)_\xi^\wedge$. The distribution character statement of Theorem 3.3.3 goes over from $G^0 \to G^0[\xi]$ to $G^0 \to G^0[\zeta]$ because the former factors through the latter. Thus we may, in addition, assume $V \subset Z$. That assumption made, $Z = V \times D$ where D has finite index in the discrete group Z_{ss}, and $\zeta = \nu \otimes \delta$ accordingly. Inclusion $G_{ss} \to G^0$ gives an isomorphism $a\colon G_{ss}[\delta] \cong G^0[\zeta]$. That gives a commutative diagram

$$\begin{array}{ccc} G_{ss}[\delta]_1^\wedge & \xrightarrow{\ \varepsilon_{ss}\ } & (G_{ss})_\delta^\wedge \\ {\scriptstyle a^*}\nearrow & & \searrow{\scriptstyle b} \\ G^0[\zeta]_1^\wedge & \xrightarrow{\qquad\varepsilon\qquad} & (G^0)_\zeta^\wedge \end{array}$$

where $b[\pi_{ss}] = [\nu \otimes \pi_{ss}]$. As in the reduction to $Z \subset G^0$, now Theorem 3.3.3 for ε_{ss} would imply the same result for ε.

We are finally reduced to the case where $Z \subset G^0$ and G^0 is semisimple. Then Z is discrete, so $S \times G^0 \to G^0[\zeta]$ is a Lie group covering. The method of reduction to simply connected G^0 now proves Theorem 3.3.3.

q.e.d.

3.4. We are going to extend the Harish-Chandra description [24] of the discrete series of a connected semisimple Lie group with finite center. In fact Harish-Chandra's analysis extends without change to connected reductive

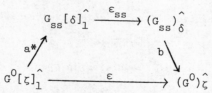

Lie groups with compact center, which is the case of G^0 where $Z \cap G^0$ is compact.

We formulate the results. Recall that G^0 is an arbitrary connected reductive Lie group and $Z \cap G^0$ is any co-compact subgroup of the center Z_{G^0} of G^0.

3.4.1. Theorem. G^0 *has a relative discrete series representation if, and only if,* $G^0/Z \cap G^0$ *has a compact Cartan subgroup.*

Let $H^0/Z \cap G^0$ be a compact Cartan subgroup(3) of $G^0/Z \cap G^0$. The Lie algebra $\mathfrak{h}$ has real dual space $\mathfrak{h}^*$, and the unitary characters on H^0 are in bijective correspondence with

(3.4.2) $L = \{\lambda \in i\mathfrak{h}^*: e^\lambda$ is well defined on $H^0\}$.

Choose a positive root system Σ^+ and make the standard definitions

(3.4.3) $\rho = \frac{1}{2} \sum_{\phi \in \Sigma^+} \phi, \quad \tilde{\omega}(\lambda) = \prod_{\phi \in \Sigma^+} \langle \phi, \lambda \rangle \quad \text{and} \quad \Delta = \prod_{\phi \in \Sigma^+} (e^{\phi/2} - e^{-\phi/2})$

where $\langle\ ,\ \rangle$ is from the Killing form. Of course $\Sigma^+ \subset L$ so $2\rho \in L$. Passing to a 2-sheeted covering group of $G^0/Z \cap G^0$, if necessary, we may assume

(3.4.4) $\rho \in L$ and $e^\rho(Z \cap G^0) = 1$.

In particular Δ is a well-defined analytic function, not only on H^0 but even on $H^0/Z \cap G^0$. Define the regular set

(3.4.5) $L' = \{\lambda \in L: \tilde{\omega}(\lambda) \neq 0\}$

and note $\rho \in L'$. If $\lambda \in L'$ we define

---

(3) The notation is justified because, if B is a compact Cartan subgroup of $G^0/Z \cap G^0$, then $B = \tilde{B}/Z \cap G^0$ where $\tilde{B}$ is a connected Cartan subgroup of G^0 that contains Z_{G^0}, hence contains $Z \cap G^0$.

(3.4.6) $q(\lambda) = |\{\text{compact } \phi \in \Sigma^+ : \langle \phi,\lambda \rangle < 0\}|$

$+ |\{\text{noncompact } \phi \in \Sigma^+ : \langle \phi,\lambda \rangle > 0\}|$,

Thus $(-1)^{q(\lambda)} = (-1)^q \operatorname{sign} \tilde{\omega}(\lambda)$ where $q = \frac{1}{2} \dim G/K$.

3.4.7. Theorem. *If* $\lambda \in L'$ *there is a unique class* $[\pi_\lambda] \in (G^0)^\wedge_{disc}$ *whose distribution character satisfies*

(3.4.8) $$\Theta_{\pi_\lambda}\Big|_{H^0 \cap G'} = (-1)^{q(\lambda)} \frac{1}{\Delta} \sum_{w \in W_{G^0}} \det(w) e^{w\lambda}$$

where W_{G^0} *is the Weyl group of* (G^0,H^0). *Every class in* $(G^0)^\wedge_{disc}$ *is one of these* $[\pi_\lambda]$. *Classes* $[\pi_\lambda] = [\pi_{\lambda'}]$ *precisely when* $\lambda' \in W_{G^0}(\lambda)$. *For appropriate normalization of Haar measure on* G^0, $[\pi_\lambda]$ *has formal degree* $|\tilde{\omega}(\lambda)|$.

3.9.4. Corollary. *The class* $[\pi_\lambda] \in (G^0)^\wedge_{disc}$ *has dual class* $[\pi^*_\lambda] = [\pi_{-\lambda}]$, *central character* $e^{\lambda-\rho}\Big|_{Z_{G^0}}$ *and infinitesimal character* χ_λ *as in* (3.2.5). *In particular* $\chi_{\pi_\lambda}(\text{Casimir}) = \|\lambda\|^2 - \|\rho\|^2$.

{The Corollary follows from formula (3.4.8).}

When Z_{G^0} is compact, Theorems 3.4.1 and 3.4.7 reduce to Harish-Chandra's celebrated results [24,Theorems 13 and 16].

Suppose that G^0 has a relative discrete series representation π. Then $[\pi] \in (G^0)^\wedge_\zeta$ for some $\zeta \in (Z \cap G^0)^\wedge$. According to Theorem 3.3.3, there is a discrete class $[\psi] \in G^0[\zeta]^\wedge_1$ such that $\varepsilon_\zeta[\psi] = [\pi]$. In particular [24,Theorem 13] $G^0[\zeta]$ has a compact Cartan subgroup. That Cartan subgroup must be an $H^0[\zeta]$ where H^0 is a Cartan subgroup of G^0. Since $H^0[\zeta]$ is compact, so is $H^0/Z \cap G^0$. This proves the "only if" part of Theorem 3.4.1.

Let $H^0/Z \cap G^0$ be a compact Cartan subgroup of $G^0/Z \cap G^0$. If

$\zeta \in (Z \cap G^0)^\wedge$ now $H^0[\zeta]$ is a compact Cartan subgroup of $G^0[\zeta]$. Write $\mathfrak{h}[\zeta]$ for its Lie algebra and define

$$L[\zeta] = \{\nu \in i\mathfrak{h}[\zeta]^*: e^\nu \text{ is well-defined on } H^0[\zeta]\}.$$

Further denote

$$L[\zeta]_1' = \{\nu \in L[\zeta]: \tilde{\omega}(\nu) \neq 0 \text{ and } e^\nu|_S = 1_S\}.$$

This will be the analog of

$$L_\zeta' = \{\lambda \in L: \tilde{\omega}(\lambda) \neq 0 \text{ and } e^\lambda|_{Z \cap G^0} = \zeta\}.$$

Since $G^0[\zeta]$ is a connected reductive Lie group with compact center, and e^ρ annihilates S, Harish-Chandra [24, Theorem 16] gives a map $\nu \to [\psi_\nu]$ of $L[\zeta]_1'$ onto $G^0[\zeta]^\wedge_{1\text{-disc}}$ which satisfies the assertions of Theorem 3.4.7. We want a similar map $\lambda \to [\pi_\lambda]$ of L_ζ' onto $(G^0)^\wedge_\zeta$. Let ω_1 and ω_ζ denote those maps; we construct ω_ζ through a commutative diagram

$$\begin{array}{ccc} L[\zeta]_1' & \xrightarrow{\ \delta\ } & L_\zeta' \\ \downarrow{\scriptstyle \omega_1} & & \downarrow{\scriptstyle \omega_\zeta} \\ G^0[\zeta]^\wedge_{1\text{-disc}} & \xrightarrow{\ \varepsilon\ } & (G^0)^\wedge_{\zeta\text{-disc}} \end{array}$$

where ε is the bijection of Theorem 3.3.3.

Let $\nu \in L[\zeta]_1'$. The distribution character of $\varepsilon\omega_1(\nu) = \omega_\zeta\delta(\nu)$ must have $(H^0 \cap G')$-restriction

$$(-1)^{q(\delta\nu)} \frac{1}{\Delta} \sum \det(w) e^{w\delta\nu} = (-1)^{q(\nu)} \frac{1}{\Delta} \sum \det(w) e^{w\nu} \cdot p.$$

For that we define δ by $e^{\delta\nu} = e^\nu \cdot p$, i.e., $\delta\nu = p^*\nu$ under $p: \mathfrak{g} \to \mathfrak{g}[\zeta]$. Since p gives an isomorphism of derived algebras of these reductive Lie algebras, now δ bijects $L[\zeta]_1'$ to L_ζ' equivariantly for W. Our assertions now go over from ω_1 to ω_ζ.

Theorems 3.4.1 and 3.4.7 are proved.

3.5. We have just worked out the relative discrete series for connected reductive Lie groups. Now we extend that to the class of reductive groups specified in §3.1. Retain the notation of §3.1.

3.5.1. Lemma. ZZ_{G^0} *has finite index in* $Z_G(G^0)$ *and every class* $[\chi] \in Z_G(G^0)^\wedge$ *has dimension* $\leqslant |Z_G(G^0)/ZZ_{G^0}|$.

Proof. Let $\tilde{Z} = ZZ_{G^0}$. According to (3.1.2), $\tilde{Z}$ is a normal abelian subgroup of some finite index r in $Z_G(G^0)$. Thus every $[\chi] \in Z_G(G^0)$ is a summand of an r-dimensional class $\mathrm{Ind}_{\tilde{Z}\uparrow Z_G(G^0)}(\beta)$, $\beta \in (\tilde{Z})^\wedge$. *q.e.d.*

3.5.2. Proposition. $\hat{G}^\dagger$ *is disjoint union of the sets*

(3.5.3) $(G^\dagger)^\wedge_\xi = \{[\chi \otimes \pi]: [\chi] \in Z_G(G^0)^\wedge_\xi$ *and* $[\pi] \in (G^0)^\wedge_\xi\}$, $\xi \in \widehat{Z_{G^0}}$.

Here $[\chi \otimes \pi]$ *has the same infinitesimal character* χ_π *as* $[\pi]$, *and* $[\chi \otimes \pi]$ *has distribution character given by the locally integrable function*

(3.5.4) $\Theta_{\chi\otimes\pi}(zg) = \mathrm{trace}\ \chi(z)\cdot\Theta_\pi(g)$ *for* $z \in Z_G(G^0)$ *and* $g \in G^0$.

Finally $[\chi \otimes \pi] \in (G^\dagger)^\wedge_{\mathrm{disc}}$ *if and only if* $[\pi] \in (G^0)^\wedge_{\mathrm{disc}}$.

Proof. Since Z_{G^0} is central in $G^\dagger$ we know (§2.2) that $\hat{G}^\dagger$ is disjoint union of the $(G^\dagger)^\wedge_\xi$, $\xi \in \widehat{Z_{G^0}}$.

Let $\xi \in \widehat{Z_{G^0}}$, $[\chi] \in Z_G(G^0)^\wedge_\xi$ and $[\pi] \in (G^0)^\wedge_\xi$. Evidently $[\chi \otimes \pi] \in (G^\dagger)^\wedge_\xi$. Further χ has some finite dimension m, so $(\chi \otimes \pi)|_{G^0} = m\pi$ where $1 \leqslant m < \infty$. Now $\chi \otimes \pi$ has infinitesimal character χ_π and is discrete relative to Z exactly when π is discrete relative to $Z \cap G^0$.

To prove (3.5.4) it suffices to consider test functions $f \in C_c^\infty(G^\dagger)$ supported in a single coset z_0G^0, $z_0 \in Z_G(G^0)$. There we compute

$$\text{trace }(\chi\otimes\pi)(f) = \text{trace}\int_{G^\dagger} f(zg)\chi(z)\otimes\pi(g)d(zg)$$

$$= \text{trace}\int_{G^0} f(z_0g)\chi(z_0)\otimes\pi(g)dg = \text{trace }\{\chi(z_0)\otimes\int_{G^0} f(z_0g)\pi(g)dg\}$$

$$= (\text{trace }\chi(z_0))(\text{trace}\int_{G^0} f(z_0g)\pi(g)dg) = (\text{trace }\chi(z_0))\int_{G^0} f(z_0g)\Theta_\pi(g)dg$$

$$= \int_{G^0} f(z_0g)(\text{trace }\chi(z_0))\Theta_\pi(g)dg = \int_{G^\dagger} f(zg)\cdot(\text{trace }\chi(z))\Theta_\pi(g)d(zg),$$

as asserted in (3.5.4).

Finally let $[\gamma] \in (G^\dagger)^\wedge_\xi$. Since $Z_G(G^0)$ acts trivially on $\hat{G}^0$ and G^0 is type I, now $\gamma|_{G^0} = m\pi$ where $1 \leqslant m \leqslant \infty$ and $[\chi] \in (G^0)^\wedge_\xi$. Similarly $\gamma|_{Z_G(G^0)} = n\chi$ where $1 \leqslant n \leqslant \infty$ and $[\chi] \in Z_G(G^0)^\wedge_\xi$. Now $[\gamma] = [\chi\otimes\pi]$ because $\chi\otimes\pi$ is a subrepresentation. That proves (3.5.3). *q.e.d.*

Proposition 3.5.2 gives us the relative discrete series of $G^\dagger$ in terms of those of $Z_G(G^0)$ and G^0. The following will let us go on to G.

3.5.5. Proposition. *Let* $[\gamma] = [\chi\otimes\pi] \in \hat{G}^\dagger$ *and define* $\psi = \text{Ind}_{G^\dagger\uparrow G}(\gamma)$.

1. $[\psi]$ *has the same infinitesimal character* χ_π *as* $[\pi]$.

2. $[\pi]$ *has distribution character which is a locally integrable function, supported in* $G^\dagger$, *given there by*

(3.5.6) $\Theta_\psi(zg) = \sum_{xG^\dagger\in G/G^\dagger} \text{trace }\chi(x^{-1}zx)\cdot\Theta_\pi(x^{-1}gx)$ *for* $z \in Z_G(G^0)$, $g \in G^0$.

In particular Θ_ψ *is analytic on the regular set* G' *and satisfies* $\Theta_\psi|_{G^\dagger} = \Theta_\psi|_{G^\dagger}$.

3. *If* $[\pi] \in (G^0)^\wedge_{disc}$ *then* $[\psi] \in \hat{G}_{disc}$. *Every class in* $\hat{G}_{disc}$ *is obtained in this way.*

Proof. We follow an argument [48, Lemma 4.3.3] of Frobenius for (1) and (2). As $G^\dagger$ is a normal subgroup of finite index in G, Θ_ψ exists and is supported in $G^\dagger$, where $\Theta_\psi|_{G^\dagger} = \Theta_{(\psi|_{G^\dagger})}$. Note $\psi|_{G^\dagger} = \sum \gamma\cdot \mathrm{ad}(x)^{-1}$ where the sum runs over the finite set $G/G^\dagger$. Now, for $z \in Z_G(G^0)$ and $g \in G^0$,

$$\Theta_\psi(zg) = \sum \Theta_{\gamma\cdot \mathrm{ad}(x)^{-1}}(zg) = \sum \Theta_\gamma(x^{-1}zx\cdot x^{-1}gx).$$

Assertion (2) now follows from (3.5.4).

Recall (3.1.1) that every $\mathrm{ad}(x)$, $x \in G$, is an inner automorphism on $\mathfrak{g}_C$. Thus $\mathrm{ad}(x)$ is trivial on the center $\mathfrak{Z}$ of the universal enveloping algebra. Now each $\gamma\cdot \mathrm{ad}(x)^{-1}$ does the same thing to $\mathfrak{Z}$. Since γ has infinitesimal character $\chi_\gamma = \chi_\pi$ and $\psi|_{G^\dagger} = \sum \gamma\cdot \mathrm{ad}(x)^{-1}$, now ψ has infinitesimal character $\chi_\psi = \chi_\pi$.

Since $|G/G^\dagger| < \infty$ every class in $\hat{G}_{disc}$ is a subrepresentation of an $[\mathrm{Ind}_{G^\dagger\uparrow G}(\gamma)]$, $[\gamma] \in (G^\dagger)^\wedge_{disc}$. If $[\gamma] = [\chi \otimes \pi]$ as in Proposition 3.5.2, the latter condition is equivalent to $[\pi] \in (G^0)^\wedge_{disc}$. To prove (3) now we need only check that $\psi = \mathrm{Ind}_{G^\dagger\uparrow G}(\gamma = \chi \otimes \pi)$ is irreducible whenever $[\pi] \in (G^0)^\wedge_{disc}$.

Choose a Cartan subgroup $H^0 \subset G^0$ with $H^0/Z \cap G^0$ compact. Then

(3.5.7a) $H = \{g \in G\colon \mathrm{ad}(g)\xi = \xi \text{ for all } \xi \in \mathfrak{h}\}$

is the corresponding Cartan subgroup of G, and

(3.5.7b) $W_G = \{g \in G\colon \mathrm{ad}(g)\mathfrak{h} = \mathfrak{h}\}/H$

is the object corresponding to the Weyl group. Our hypothesis (3.1.1) says that W_G is a subgroup of the complex Weyl group $W(\mathfrak{g}_C, \mathfrak{h}_C)$. Since any two compact Cartan subgroups of $G^0/Z \cap G^0$ are conjugate, we have a system $\{x_1, \ldots, x_r\}$ of representatives of G modulo $G^\dagger$ such that each $\mathrm{ad}(x_j)\mathfrak{h} = \mathfrak{h}$. Thus

(3.5.7c) $W_G = \bigcup_{1\leqslant j\leqslant r} (x_j H) W_{G^0} \subset W(\mathfrak{g}_C, \mathfrak{h}_C)$.

Suppose $[\gamma] = [\chi \otimes \pi]$ with $[\pi] \in (G^0)^\wedge_{disc}$. Use the terminology of Theorem 3.4.7. Choose $\lambda \in L' \subset i\mathfrak{h}^*$ such that $[\pi] = [\pi_\lambda]$. Then $[\pi \cdot ad(x_j)^{-1}] = [\pi_{\lambda_j}]$ where $\lambda_j = ad(x_j^{-1})^*(\lambda)$. Since $\lambda \in L'$ the $w(\lambda)$, $w \in W(\mathfrak{g}_C, \mathfrak{h}_C)$, are distinct. Now (3.5.7c) shows the λ_j distinct modulo the action of W_{G^0}. Theorem 3.4.7 now says that the $\pi \cdot ad(x_j)^{-1}$ are mutually inequivalent. Thus also the $\gamma \cdot ad(x_j)^{-1} = (\chi \cdot ad(x_j)^{-1} \otimes (\pi \cdot ad(x_j)^{-1})$ are mutually inequivalent. Since $\psi = Ind_{G^\dagger \uparrow G}(\gamma)$ has $\psi|_{G^\dagger} = \sum \gamma \cdot ad(x_j)^{-1}$, now ψ is irreducible. *q.e.d.*

We formulate the extensions of Theorems 3.4.1 and 3.4.7 from G^0 to G.

3.5.8. Theorem. *G has a relative discrete series representation if and only if G/Z has a compact Cartan subgroup.*

Let H/Z be a compact Cartan subgroup of G/Z. Then $H^0/Z \cap G^0$ is a compact Cartan subgroup of $G^0/Z \cap G^0$, and we retain the notation (3.4.2)-(3.4.6) used to describe $(G^0)^\wedge_{disc}$. Also retain the notation involved in (3.5.7): $G = \bigcup_{1 \leq j \leq r} x_j G^\dagger$ where x_j normalizes H^0. Denote

$$w_j: \text{ element of } W_G \text{ represented by } x_j,$$

so $w_j(\nu) = ad(x_j^{-1})^* \nu$ for all $\nu \in \mathfrak{h}_C^*$.

3.5.9. Theorem. *Let $\lambda \in L'$ and $[\chi] \in Z_G(G^0)^\wedge_\xi$ where $\xi = e^{\lambda-\rho}|_{Z_{G^0}}$. If $[\pi_\lambda] \in (G^0)^\wedge_{disc}$ is the class specified in Theorem 3.4.7, then*

(3.5.10a) $[\pi_{\chi,\lambda}] = [Ind_{G^\dagger \uparrow G}(\chi \otimes \pi_\lambda)]$

is the unique class in $\hat{G}_{disc}$ whose distribution character satisfies

(3.5.10b) $\Theta_{\pi_{\chi,\lambda}}(zh) = \sum_{1 \leq j \leq r} (-1)^{q(w_j\lambda)} \, tr\, \chi(x_j^{-1} z x_j) \cdot \frac{1}{\Delta} \sum_{w \in W_{G^0}} \det(ww_j) e^{ww_j\lambda}(h)$

for $z \in Z_G(G^0)$ and $h \in H^0 \cap G'$.

Every class in $\hat{G}_{disc}$ is one of the $[\pi_{\chi,\lambda}]$ just specified. Classes

$[\pi_{\chi,\lambda}] = [\pi_{\chi',\lambda'}]$ *precisely when* $([\chi'],\lambda') \in W_G([\chi],\lambda)$. *For appropriate normalization of Haar measures, the formal degree*

$\deg(\pi_{\chi,\lambda}) = r\cdot\dim(\chi)\cdot|\tilde{\omega}(\lambda)|$.

The class $[\pi_{\chi,\lambda}]$ *has dual* $[\pi_{\chi^*,-\lambda}]$ *and has infinitesimal character* χ_λ *as in* (3.2.5). *In particular* $\chi_{\pi_{\chi,\lambda}}(\text{Casimir}) = \|\lambda\|^2 - \|\rho\|^2$.

Proof. If $\hat{G}_{disc}$ is non-empty, Propositions 3.5.2 and 3.5.5 show $(G^0)^\wedge_{disc}$ non-empty, so $G^0/Z \cap G^0$ has a compact Cartan subgroup by Theorem 3.4.1. As $G^0/Z \cap G^0$ has finite index in G/Z, the latter also has a compact Cartan subgroup.

If H/Z is a compact Cartan subgroup of G/Z, then Theorem 3.4.7 and Propositions 3.5.2 and 3.5.5 force Theorem 3.5.9. *q.e.d.*

§4. Nondegenerate Series Representations of Reductive Lie Groups.

G is a reductive Lie group of the class described in §3.1. If $\{gHg^{-1}\}_{g \in G}$ is a conjugacy class of Cartan subgroups of G with H/Z compact, then in §3 we used H to construct the relative discrete series $\hat{G}_{disc} \subset \hat{G}$. Here we take an arbitrary conjugacy class of Cartan subgroups of G and use it to construct a series of unitary representations.

In §§4.1 and 4.2 we work out the relations between Cartan involutions θ of G, Cartan subgroups H of G, and cuspidal parabolic subgroups $P = MAN$ of G. Here H has a canonical splitting as $T \times A$ where T/Z is compact and A is split/$\mathbb{R}$. The G-centralizer of A has canonical splitting as $M \times A$ where M is in the class of §3.1 and T is a Cartan subgroup of M. Then $H \mapsto P$ induces a bijection from the conjugacy classes of Cartan subgroups to the "association classes" of cuspidal parabolic subgroups of G.

In §4.3 we describe the unitarily induced representations of G specified by an association class of cuspidal parabolic subgroups $P = MAN$. Let $[\eta] \in \hat{M}$, $e^{i\sigma} \in \hat{A}$, and $\pi_{\eta,\sigma} = \mathrm{Ind}_{P\uparrow G}(\eta \otimes e^{i\sigma})$ where $(\eta \otimes e^{i\sigma})(man) = e^{i\sigma}(a)\eta(m)$. In Theorem 4.3.8 we show that $[\pi_{\eta,\sigma}]$ is a finite sum from $\hat{G}$, and we completely calculate its infinitesimal and distribution characters in terms of $[\eta]$ and σ. In particular, $[\pi_{\eta,\sigma}]$ depends only on $[\eta \otimes e^{i\sigma}] \in (M \times A)^{\wedge}$, rather than the choice of N in P and the corresponding class in $\hat{P}$.

In §4.4 we specialize to the case $[\eta] \in \hat{M}_{disc}$. The corresponding $[\pi_{\eta,\sigma}]$ form the "H-series" of G. Parameterizing $\hat{M}_{disc}$ as the $[\eta_{\chi,\nu}]$ in the manner of §3.5, the H-series consists of the $[\pi_{\chi,\nu,\sigma}] = [\pi_{\eta_{\chi,\nu},\sigma}]$, and $[\pi_{\chi,\nu,\sigma}]$ depends precisely on the Weyl group orbit $W_{G,H}([\chi],\nu,\sigma)$. Finally we show that representations from different series are disjoint.

In §4.5 we state an irreducibility theorem of Harish-Chandra. That irreducibility is not needed in the sequel, but it puts the results of §§4.3 and 4.4 in their proper perspective. I am indebted to Harish-Chandra for showing me his manuscript.

The general ideas used in §4 have been known for many years, at least for connected semisimple groups with finite center. Using methods of Harish-Chandra, and pushing them when necessary, we work out the technical details needed for the class of reductive groups specified in §3.1.

4.1. G is a reductive Lie group that satisfies the conditions (3.1.1) and (3.1.2). We retain the notation of §3.1.

Let $\mathfrak{h}$ be a Cartan subalgebra of $\mathfrak{g}$. The associated Cartan subgroup is its G-centralizer

$$H = \{g \in G: \mathrm{ad}(g)\xi = \xi \text{ for all } \xi \in \mathfrak{h}\}.$$

If G^0 is a linear group, or if H/Z is compact, then $H \cap G^0$ is commutative. In general one only knows that $H^0 = \exp(\mathfrak{h})$ is commutative.

4.1.1. Lemma. *If K/Z is a maximal compact subgroup of G/Z, then G has a unique involutive automorphism θ with fixed point set K.*

These automorphisms are called Cartan involutions of G.

4.1.2. Lemma. *If K/Z is a maximal compact subgroup of G/Z, then $K^0 = K \cap G^0$, K meets every component of G, and $K = \{g \in G: \mathrm{ad}(g)K^0 = K^0\}$.*

Any two Cartan involutions of G are conjugate by an element of $\mathrm{ad}(G^0)$.

Every Cartan subgroup of G is stable under a Cartan involution.

The assertions of Lemmas 4.1.1 and 4.1.2 are standard in the cases where $Z = \{1\}$ and G is either linear or semisimple.

Proof of Lemmas. K contains Z_{G^0} because (3.1.2c) KZ_{G^0}/Z is compact. As $(K \cap G^0)/(Z_{G^0})^0$ is connected, is its own $G^0/(Z_{G^0})^0$-normalizer, and is unique up to conjugacy, the same holds for $K \cap G^0$ in G^0. Let $E = \{g \in G\colon \mathrm{ad}(g)K^0 = K^0\}$; now $E \cap G^0 = K^0 = K \cap G^0$. If $g \in G$ some $g' \in G^0$ has $\mathrm{ad}(g')(\mathrm{ad}(g)K^0) = K^0$, so E meets gG^0. Now $K \subset E \subset K$ and the first assertion of Lemma 4.1.2 is proved.

From symmetric space theory, each simple ideal $\mathfrak{g}_i$ of $\mathfrak{g}$ has a unique involutive automorphism θ_i with fixed point set $\mathfrak{k} \cap \mathfrak{g}_i$. Direct sum of the θ_i and the identity map of the center of $\mathfrak{g}$, gives the unique involutive automorphism θ of $\mathfrak{g}$ with fixed point set $\mathfrak{k}$. Extend θ to the universal cover of G^0; there its fixed point set $\exp(\mathfrak{k})$ contains the center; so θ extends uniquely to G^0 with fixed point set K^0. Now θ extends uniquely to $G = KG^0$ with fixed point set K, by the first assertion of Lemma 4.1.2, using the formula $\theta(kg) = k\theta(g)$ for $k \in K$ and $g \in G^0$. Lemma 4.1.1 is proved.

Any two choices of K/Z are $(G/Z)^0$-conjugate. Thus any two choices of $\mathfrak{k}$ are G^0-conjugate. By the first assertion of Lemma 4.1.2 now any two choices of K are G^0-conjugate. Now Lemma 4.1.1 gives the second assertion of Lemma 4.1.2. The third assertion follows from the corresponding fact about Cartan subalgebras. *q.e.d.*

Fix the data

(4.1.3a) H: Cartan subgroup of G,

(4.1.3b) θ: Cartan involution of G with $\theta(H) = H$, and

(4.1.3c) K: fixed point set of θ on G.

We decompose

(4.1.4a) $\mathfrak{h} = \mathfrak{t} + \mathfrak{a}$ into (+1)- and (-1)-eigenspaces of $\theta|_{\mathfrak{h}}$.

That splits H as direct product

(4.1.4b) $H = T \times A$ where $T = H \cap K$ and $A = \exp(\mathfrak{a})$.

The $\mathfrak{a}$-roots of $\mathfrak{g}$ are the nonzero real linear functionals $\phi \in \mathfrak{a}^*$ such that

$$\mathfrak{g}^{\phi} = \{\xi \in \mathfrak{g} : [\alpha,\xi] = \phi(\alpha)\xi \text{ for all } \alpha \in \mathfrak{a}\}$$

is nonzero. Let $\Sigma_{\mathfrak{a}}$ denote the set of $\mathfrak{a}$-roots of $\mathfrak{g}$. Then we have an $\mathfrak{a}$-root space decomposition

$$\mathfrak{g} = \mathfrak{z}_{\mathfrak{g}}(\mathfrak{a}) + \sum_{\phi\in\Sigma_{\mathfrak{a}}} \mathfrak{g}^{\phi} \quad \text{where} \quad \mathfrak{z}_{\mathfrak{g}}(\mathfrak{a}) = \{\xi \in \mathfrak{g} : [\xi,\mathfrak{a}] = 0\}.$$

This refines as follows.

4.1.5. Lemma. *The* G-*centralizer of* A *has unique splitting* $Z_G(A) = M \times A$ *with* $\theta(M) = M$. *In particular* $\mathfrak{g}$ *has unique decomposition* $\mathfrak{m} + \mathfrak{a} + \sum_{\phi\in\Sigma_{\mathfrak{a}}} \mathfrak{g}^{\phi}$ *with* $\theta(\mathfrak{m}) = \mathfrak{m}$.

Proof. Necessarily $\mathfrak{m}$ is the sum of $\mathfrak{z}_{\mathfrak{g}}(\mathfrak{a}) \cap \mathfrak{k}$ with the nontrivial $\mathrm{ad}\{\mathfrak{z}_{\mathfrak{g}}(\mathfrak{a}) \cap \mathfrak{k}\}$-modules on $\mathfrak{z}_{\mathfrak{g}}(\mathfrak{a})$. By Lemma 4.1.2 on $Z_G(A)$, K meets every component of $Z_G(A)$. Let M^0 be the analytic subgroup of G for $\mathfrak{m}$. Necessarily $M = (K \cap Z_G(A))M^0$.

q.e.d.

We will keep the notation $Z_G(A) = M \times A$.

The hereditary property of the conditions of §3.1 is

4.1.6. Proposition. M *inherits the conditions of* §3.1 *from* G: *every* $\mathrm{ad}(m)$ *is inner on* $\mathfrak{m}_{\mathbb{C}}$, Z *centralizes* M^0, $|M/ZM^0| < \infty$ *and* $Z \cap M^0$ *is co-compact in the center of* M^0. *Further*, T/Z *is a compact Cartan subgroup of* M/Z.

Before proving the Proposition it is convenient to introduce positive $\mathfrak{a}$-root systems.

Every $\phi \in \Sigma_{\mathfrak{a}}$ specifies a hyperplane $\phi^{\perp} = \{\alpha \in \mathfrak{a} : \phi(\alpha) = 0\}$. The complement $\mathfrak{a} - \bigcup_{\Sigma_{\mathfrak{a}}} \phi^{\perp}$ is a finite union of convex open cones, its topological components; those cones are called Weyl chambers. Choice of a Weyl chamber $\mathfrak{d} \subset \mathfrak{a}$ specifies a positive $\mathfrak{a}$-root system $\Sigma_{\mathfrak{a}}^{+} = \{\phi \in \Sigma_{\mathfrak{a}}: \phi$ takes positive values in $\mathfrak{d}\}$.

4.1.7. Lemma. *Fix a positive $\mathfrak{a}$-root system $\Sigma_{\mathfrak{a}}^{+}$ on $\mathfrak{g}$ and a positive $\mathfrak{t}_{C}$-root system $\Sigma_{\mathfrak{t}}^{+}$ on $\mathfrak{m}_{C}$. Then there is a unique positive $\mathfrak{h}_{C}$-root system Σ^{+} on $\mathfrak{g}_{C}$ such that*

$$\Sigma_{\mathfrak{a}}^{+} = \{\gamma|_{\mathfrak{a}} : \gamma \in \Sigma^{+} \text{ and } \gamma|_{\mathfrak{a}} \neq 0\}$$

and

$$\Sigma_{\mathfrak{t}}^{+} = \{\gamma|_{\mathfrak{t}} : \gamma \in \Sigma^{+} \text{ and } \gamma|_{\mathfrak{a}} = 0\}.$$

Proof. Choose ordered bases $\beta_{\mathfrak{a}}$ of $\mathfrak{a}^{*}$ and $\beta_{\mathfrak{t}}$ of $i\mathfrak{t}^{*}$, whose associated lexicographic orderings yield $\Sigma_{\mathfrak{a}}^{+}$ and $\Sigma_{\mathfrak{t}}^{+}$ as respective positive root systems. Then the ordered basis $\beta = \{\beta_{\mathfrak{a}}, \beta_{\mathfrak{t}}\}$ of $\mathfrak{a}^{*} + i\mathfrak{t}^{*}$ gives a lexicographic order, whose associated positive $\mathfrak{h}_{C}$-root system Σ^{+} has the required properties.

Let Σ^{+} and $'\Sigma^{+}$ be positive $\mathfrak{h}_{C}$-root systems on $\mathfrak{g}_{C}$ that yield $\Sigma_{\mathfrak{a}}^{+}$ and $\Sigma_{\mathfrak{t}}^{+}$ as specified. Suppose $\gamma \in \Sigma^{+}$. If $\gamma|_{\mathfrak{a}} \neq 0$ then $\gamma|_{\mathfrak{a}} \in \Sigma_{\mathfrak{a}}^{+}$ so $\gamma \in {'\Sigma^{+}}$. If $\gamma|_{\mathfrak{a}} = 0$ then $\gamma|_{\mathfrak{t}} \in \Sigma_{\mathfrak{t}}^{+}$ so $\gamma \in {'\Sigma^{+}}$. Now $'\Sigma^{+} \subset \Sigma^{+}$. Similarly $\Sigma^{+} \subset {'\Sigma^{+}}$. Uniqueness is proved. *q.e.d.*

Proof of Proposition 4.1.6. Since $H \subset M \times A$ and H is a Cartan subgroup of G, also H is a Cartan subgroup of $M \times A$. As $T \subset M$ now T is a Cartan subgroup of M. Since K/Z is compact, so is T/Z.

Let $m \in M$ such that $ad(m)$ is an outer automorphism of $\mathfrak{m}_{C}$. We may move m within mM^{0} and assume $ad(m)\mathfrak{t} = \mathfrak{t}$. Then $ad(m)$ preserves, but acts nontrivially on, a positive $\mathfrak{t}_{C}$-root system $\Sigma_{\mathfrak{t}}^{+}$ of $\mathfrak{m}_{C}$; see [48, Theorem 8.11.2]. Choose a positive $\mathfrak{a}$-root system $\Sigma_{\mathfrak{a}}^{+}$ on $\mathfrak{g}$; $ad(m)$ preserves and acts trivially on $\Sigma^+_{\mathfrak{a}}$ because m centralizes $\mathfrak{a}$. Let Σ^+ be the positive $\mathfrak{h}_C$-root system on $\mathfrak{g}_C$ determined by $\Sigma^+_{\mathfrak{a}}$ and $\Sigma^+_{\mathfrak{t}}$ according to Lemma 4.1.7. Now $ad(m)$ preserves Σ^+ by uniqueness. Since $ad(m)$ is inner on $\mathfrak{g}_C$ (3.1.1) it follows that $ad(m)$ is trivial on Σ^+. That contradicts non-triviality of $ad(m)$ on $\Sigma^+_{\mathfrak{t}}$. We conclude that every $ad(m)$, $m \in M$, is an inner automorphism on $\mathfrak{m}_C$.

We check $|M/ZM^0| < \infty$. From the two facts proved so far, $M^\dagger = \{m \in M: ad(m) \text{ inner on } M^0\}$ has finite index on M and $M^\dagger = TM^0$. Thus T/Z is a compact subgroup of M/Z such that the coset space $M/T = (M/Z)/(T/Z)$ has only finitely many components. Now M/Z has only a finite number of components, i.e. $|M/ZM^0| < \infty$.

The center of $M^0/Z \cap M^0$ is a closed subgroup of the torus $T^0/Z \cap M^0$, hence compact. That center is $Z_{M^0}/Z \cap M^0$. *q.e.d.*

4.2. We apply the considerations of §4.1 to study the cuspidal parabolic subgroups of G. Retain the data (4.1.3) and the splittings $H = T \times A$ and $Z_G(A) = M \times A$.

Choose a positive $\mathfrak{a}$-root system $\Sigma^+_{\mathfrak{a}}$ on $\mathfrak{g}$. It defines

(4.2.1a) $\mathfrak{n} = \sum_{\phi \in \Sigma^+_{\mathfrak{a}}} \mathfrak{g}^{-\phi}$ nilpotent subalgebra of $\mathfrak{g}$,

(4.2.1b) $N = \exp(\mathfrak{n})$ unipotent analytic subgroup of G, and

(4.2.1c) $P = \{g \in G: ad(g)N = N\}$ normalizer of N in G.

4.2.2. Lemma. P *is a (real) parabolic subgroup of* G. *It has unipotent radical* $P^u = N$ *and has reductive part* $P^r = M \times A$. *Thus* $P = MAN$ *in the sense of smooth unique factorization.*

{This result is standard in the case where $G = G^\dagger$ and G is linear.}

Proof. Let $\pi: G \to \bar{G}$ denote projection to the real linear algebraic group $\bar{G} = G/ZZ_{G^0}$. Then $\pi(P)$ is the parabolic subgroup of $\bar{G}$ which is normalizer of $\pi(N)$; and $\pi(N)$ is the unipotent radical $\pi(P)^u$. If $\pi(P)^r$ is any maximal reductive algebraic subgroup of $\pi(P)$ then we have $\pi(P) = \pi(P)^r \cdot \pi(N)$ semidirect product.

Note $ZZ_{G^0} \subset MA \subset P$. Thus (i) we can choose $\pi(P)^r$ to contain $\pi(MA)$ and (ii) $P = P^r \cdot N$ semidirect product where $P^r = \pi^{-1}(\pi(P)^r)$. Now $MA \subset P^r$, and a dimension count shows that M^0A is the identity component of P^r. The uniqueness in Lemma 4.1.5 says that A is normal in P^r.

We choose an element $x \in \mathfrak{a}$ that is moved by every element w of the complex Weyl group $W(\mathfrak{g}_C, \mathfrak{h}_C)$ such that $w(\mathfrak{a}) = \mathfrak{a}$ but $w|_{\mathfrak{a}}$ is nontrivial. If V is the set of all such w, then we simply choose x in the complement of the finite union $\bigcup_{w \in V} \{y \in \mathfrak{a}: w(y) = y\}$ of lower-dimensional subspaces of $\mathfrak{a}$.

Let $g \in P^r$. Since $ad(g)$ is inner on $\mathfrak{g}_C$ now x and $ad(g)x$ are conjugate by an inner automorphism of $\mathfrak{g}_C$. Thus [42, Theorem 2.1] they are conjugate by an inner automorphism of $\mathfrak{g}$. Now some $g' \in gG^0$ has $ad(g')x = x$. Since x and $\mathfrak{a}$ have the same $\mathfrak{g}$-centralizer it follows from Lemma 4.1.5 that $ad(g')M = M$. Thus we may assume $ad(g')H = H$. By choice of x now g' centralizes $\mathfrak{a}$, so $g' \in MA$. We have proved that $P^r = MA(P^r \cap G^0)$. However $P^r \cap G^0$ centralizes A. Thus $P^r = MA$ as required. *q.e.d.*

We say that two parabolic subgroups of G are <u>associated</u> if their reductive parts are G-conjugate. As $\Sigma^+_{\mathfrak{a}}$ ranges over the positive $\mathfrak{a}$-root systems of $\mathfrak{g}$, the parabolic subgroup P of (4.2.1) remains in the same association class.

Let Q be a parabolic subgroup of G. We say that Q is <u>cuspidal</u> if the derived group of $(Q^r)^0$ has a Cartan subgroup E such that $E/E \cap Z_{G^0}$ is compact.

4.2.3. Proposition. *If* Q *is a parabolic subgroup of* G *then the following conditions are equivalent.*

(i) Q *is a cuspidal parabolic subgroup of* G.

(ii) *There is a Cartan subgroup* $H = T \times A$ *of* G, *and a positive* $\mathfrak{a}$-*root system* $\Sigma^+_{\mathfrak{a}}$, *such that* Q *is the group* P *of* (4.2.1).

(iii) Q^r *has a relative discrete series representation.*

(iv) $(Q^r)^0$ *has a relative discrete series representation.*

In particular, the construction $H \mapsto P = MAN$ *of* (4.2.1) *induces a bijection from the set of all conjugacy classes of Cartan subgroups of* G *onto the set of all association classes of cuspidal parabolic subgroups of* G.

Proof. Let $\pi: G \to \bar{G} = G/ZZ_{G^0}$ as before and observe $ZZ_{G^0} \subset Q^r \subset Q$. Thus each of conditions (i), (ii) and (iii) holds for Q exactly when it holds for $\pi(Q)$. On the other hand, (iii) and (iv) are equivalent because Q^r/ZZ_{G^0} has only finitely many topological components. For the equivalence now, it suffices to prove (i), (ii) and (iv) equivalent when G^0 is centerless and G/G^0 is finite. And there it evidently suffices to consider the case where G is connected.

Now G is a connected centerless semisimple group. So $Q^r = M_Q \times A_Q$ where A_Q is the R-split component of the center of Q^r, and both M_Q and A_Q are stable under a Cartan involution θ. Thus Q is cuspidal if and only if M_Q has a compact Cartan subgroup T_Q. That is the case precisely when G has a Cartan subgroup $H = T_Q \times A_Q$ from which Q is constructed by (4.2.1). This proves equivalence of (i) and (ii). Apply Theorem 3.4.1 to $(Q^r)^0$. Then (ii) implies (iv) by Proposition 4.1.6 and (iv) implies (i) directly.

We have proved equivalence of (i), (ii), (iii) and (iv). The remaining statement follows from Lemmas 4.1.5 and 4.2.2. *q.e.d.*

4.2.4. Remark. L. P. Rothschild showed that two Cartan subalgebras of $\mathfrak{g}$ are conjugate by an inner automorphism of $\mathfrak{g}$ precisely when they are conjugate by an inner automorphism of $\mathfrak{g}_C$ [42, Corollary 2.4]. In view of (3.1.1), now conjugacy classes of Cartan subgroups of G are the same as G^0-conjugacy classes. In view of Proposition 4.2.3 the analogous fact holds for association classes of cuspidal parabolic subgroups of G. Thus we could use G^0-conjugacy, G^0-association, or both, in the bijection of Proposition 4.2.3.

4.3. We define the series of unitary representations of G associated to a conjugacy class of Cartan subgroups. Then we work out the character theory for that series in a general way. The precise character theory is in §4.4.

Retain the notation of §§4.1 and 4.2, especially (4.1.3) and the splittings $H = T \times A$, $Z_G(A) = M \times A$ and $P = MAN$.

The general class in $(P^r)^\wedge = (M \times A)^\wedge$ is an

(4.3.1a) $[\eta \otimes e^{i\sigma}] \in (P^r)^\wedge$ where $[\eta] \in \hat{M}$ and $\sigma \in \mathfrak{a}^*$.

That extends to a class in $\hat{P}$ which annihilates N:

(4.3.1b) $[\eta \otimes e^{i\sigma}] \in \hat{P}$ by $(\eta \otimes e^{i\sigma})(man) = e^{i\sigma}(a)\eta(m)$.

Now we have the representation of G given by

(4.3.1c) $\pi_{\eta,\sigma} = \mathrm{Ind}_{P\uparrow G}(\eta \otimes e^{i\sigma})$ unitarily induced representation.

By the H-series of G we mean $\{[\pi_{\eta,\sigma}]: \eta \in \hat{M}_{disc}$ and $\sigma \in \mathfrak{a}^*\}$. If H/Z is compact then $M = G$ and the H-series is just the relative discrete series $\hat{G}_{disc}$. If H/ZZ_{G^0} is maximally R-split, i.e. if P is a minimal parabolic subgroup of G, then M/Z is compact, so $\hat{M} = \hat{M}_{disc}$; then the H-series is usually called the principal series of G. We will refer to any H-series as a nondegenerate series of representations of G. Unfortunately there is no really standard terminology here.

Later we will see that the H-series depends only on the conjugacy class of H (or, equivalently, the association class of P) rather than the conjugacy class of P.

Recall that P was defined (4.2.1) from a choice $\Sigma_{\mathfrak{a}}^{+}$ of positive $\mathfrak{a}$-root system on $\mathfrak{g}$. Define

(4.3.2a) $\rho_{\mathfrak{a}} = \frac{1}{2}\sum_{\phi\in\Sigma_{\mathfrak{a}}^{+}} (\dim \mathfrak{g}^{\phi})\phi$

so that, under the adjoint representation of $\mathfrak{g}$,

(4.3.2b) $\mathfrak{a}$ acts on $\mathfrak{n}$ and $\mathfrak{p}$ with trace -2ρ.

Then $P = MAN$ has modular function δ_P,

$$\int_P f(xy^{-1})dx = \delta_P(y)\int_P f(x)dx,$$

given by

(4.3.2c) $\delta_P(man) = e^{2\rho_{\mathfrak{a}}}(a)$ for $m \in M$, $a \in A$, $n \in N$.

Let $[\eta] \in \hat{M}$ and let E_η denote its representation space. Given $\sigma \in \mathfrak{a}^*$ we have a Hilbert space bundle

(4.3.3a) $\mathcal{E}_{\eta,\sigma} \to G/P = K/K \cap M$

associated to the non-unitary representation $\eta \otimes e^{\rho_{\mathfrak{a}}+i\sigma}$ of P. Here G acts on the bundle but the hermitian metric is stable only under K; we are using $G = KP$ and $K \cap P = K \cap M$. The K-invariant probability measure on G/P is specified by normalized Haar measure on K/Z. Now the space of square integrable sections of $\mathcal{E}_{\eta,\sigma} \to G/P$ is

$$(4.3.3b)\quad L_2(\mathcal{E}_{\eta,\sigma}):\ \begin{cases} \text{all Borel-measurable } f\colon G \to E_\eta \text{ such that} \\ \text{(i)}\ \ f(gp) = (\eta \otimes e^{\rho_{\mathfrak{a}}+i\sigma})(p)^{-1} f(g) \text{ and} \\ \text{(ii)}\ \ \int_{K/Z} \langle f(k), f(k)\rangle\, d(kZ) < \infty \end{cases}$$

where we identify functions that agree a.e. on G. $L_2(\mathcal{E}_{\eta,\sigma})$ is a Hilbert space with inner product

$$\langle f, f' \rangle = \int_{K/Z} \langle f(k), f'(k)\rangle\, d(kZ).$$

G acts on $L_2(\mathcal{E}_{\eta,\sigma})$ by the "left regular" representation $[g(f)](g') = f(g^{-1}g')$. In view of (4.3.2c), unimodularity of G and the definition of the unitarily induced representation (4.3.1c) give

(4.3.3c) $\pi_{\eta,\sigma}$ is the (left regular) representation of G on $L_2(\mathcal{E}_{\eta,\sigma})$.

Now we are going to work out the structure of the representations $\pi_{\eta,\sigma}$. We first collect some preliminaries on Cartan subgroups and regular elements. $H = T \times A$ and $P = MAN$ remain fixed as above.

If J is a Cartan subgroup of G, then these conditions are equivalent: (i) $J \subset MA$, (ii) J is a Cartan subgroup of MA, (iii) $J = J_M \times A$ where $J_M = J \cap M$ is a Cartan subgroup of M.

Let $J = J_M \times A$, Cartan subgroup of G that is contained in MA. Without loss of generality we may assume J_M stable under the Cartan involution $\theta|_M$ of M.

Choose a positive $(\mathfrak{j}_M)_{\mathbb{C}}$-root system $\Sigma^+_{\mathfrak{j}_M}$ on $\mathfrak{m}_{\mathbb{C}}$. As in Lemma 4.1.7, there is a unique positive $\mathfrak{j}_{\mathbb{C}}$-root system $\Sigma^+_{\mathfrak{j}}$ on $\mathfrak{g}_{\mathbb{C}}$ such that

(4.3.4a) $\Sigma^+_{\mathfrak{a}} = \{\gamma|_{\mathfrak{a}} : \gamma \in \Sigma^+ \text{ and } \gamma|_{\mathfrak{a}} \neq 0\}$ and $\Sigma^+_{\mathfrak{j}_M} = \{\gamma|_{\mathfrak{j}_M} : \gamma \in \Sigma^+ \text{ and } \gamma|_{\mathfrak{a}} = 0\}$.

Then let

(4.3.4b) $\rho_{\mathfrak{j}} = \frac{1}{2}\sum_{\gamma\in\Sigma^+_{\mathfrak{j}}} \gamma$ so $\rho_{\mathfrak{a}} = \rho_{\mathfrak{j}}|_{\mathfrak{a}}$

and define

(4.3.4c) $\Delta_{G,J} = \prod_{\gamma\in\Sigma_j^+} (e^{\gamma/2} - e^{-\gamma/2})$ and $\Delta_{M,J_M} = \prod_{\phi\in\Sigma_{j_M}^+} (e^{\phi/2} - e^{-\phi/2})$.

4.3.5. Lemma. *If* γ *is a* j_C*-root, then* e^γ *is a well-defined character on* J, *unitary on* $J \cap K$, *whose kernel contains* $Z_G(G^0)$. *If* ϕ *is a* $(j_M)_C$*-root, say* $\phi = \gamma|_{j_M}$ *where* γ *is a* j_C*-root, then the corresponding character on* J_M *is* $e^\phi = e^\gamma|_{J\cap M}$.

4.3.6. Lemma. *We can cut* Z *down to a subgroup of index* $\leqslant 2$, *or replace* G *by a* Z_2*-extension, so that the following condition holds. If* L *is any Cartan subgroup of* G *and we make any choice of positive* $\mathfrak{l}_C$*-root system on* $\mathfrak{g}_C$, *then*

$$e^{\rho_\mathfrak{l}} \quad \textit{is a well-defined character on } L \textit{ with } e^{\rho_\mathfrak{l}}(Z) = 1$$

and

$$\Delta_{G,L} \quad \textit{is a well-defined analytic function on } L.$$

In particular then e^{ρ_j} *and* $\Delta_{G,J}$ *are well defined on* J, *and it follows that* $e^{\rho_{j_M}}$ *and* Δ_{M,J_M} *are well defined on* J_M.

Proof of Lemmas. The inner automorphisms of $\mathfrak{g}_C$ form a connected centerless complex Lie group $\mathrm{Int}(\mathfrak{g}_C)$. By (3.1.1), $\mathrm{ad}_\mathfrak{g}$ maps G to a (possibly disconnected) real form $G/Z_G(G^0)$ of $\mathrm{Int}(\mathfrak{g}_C)$. If γ is a j_C-root now e^γ is well defined on the Cartan subgroup $(J/Z_G(G^0))_C$ of $\mathrm{Int}(\mathfrak{g}_C)$, for that Cartan subgroup is connected. For Lemma 4.3.5 one restricts e^γ to $J/Z_G(G^0)$ and lifts that to J. This argument gives us a 1- or 2-sheeted covering $q: Q \to \mathrm{Int}(\mathfrak{g}_C)$ such that e^{ρ_j} is well defined on $q^{-1}(J/Z_G(G^0))_C$.

Cutting Z to a subgroup of index $\leqslant 2$, or perhaps replacing G by a Z_2-extension, we factor $\mathrm{ad}_\mathfrak{g}: G \to \mathrm{Int}(\mathfrak{g}_C)$ as $G \to G/Z \to Q \to \mathrm{Int}(\mathfrak{g}_C)$. That done, e^{ρ_j} is well defined on J and has Z in its kernel. Recall L

and $\rho_{\mathfrak{l}}$ from Lemma 4.3.6. There exists $\bar{x} \in \mathrm{Int}(\mathfrak{g}_C)$ such that $\mathrm{ad}(\bar{x})(L/Z_G(G^0))_C = (J/Z_G(G^0))_C$ and $\mathrm{ad}(\bar{x})^*\rho_{\mathfrak{j}} = \rho_{\mathfrak{l}}$. Now $e^{\rho_{\mathfrak{l}}}$ is well defined on $q^{-1}(L/Z_G(G^0))_C$, thus on L with Z in its kernel. So also $\Delta_{G,L} = e^{-\rho_{\mathfrak{l}}} \cdot \prod_{\Sigma_{\mathfrak{l}}^+} (e^{\gamma}-1)$ is well defined on L.

Q was defined so that it has a faithful irreducible holomorphic representation σ whose highest weight relative to $(\mathfrak{j}, \Sigma_{\mathfrak{j}}^+)$ is $\rho_{\mathfrak{j}}$. We recognize that as a spin representation: Let $C\ell(G)$ denote the Clifford algebra over the Lie algebra of $\mathrm{Int}(\mathfrak{g}_C)$ relative to the Killing form and $C\ell^*(G)$ its multiplicative group of units. Then $Q \subset C\ell^*(G)$ and σ is a summand of the left multiplication representation λ_Q of Q on $C\ell(G)$. The Clifford algebra $C\ell(M) \subset C\ell(G)$ and is stable under $\lambda_Q(q^{-1}(\mathrm{ad}_{\mathfrak{g}}(M)))$. In the representation of M on $C\ell(M)$ we have an irreducible summand of highest weight $\rho_{\mathfrak{j}_M}$. Now $e^{\rho_{\mathfrak{j}_M}}$ and Δ_{M,J_M} are well defined on J_M. *q.e.d.*

We are using G' to denote the regular set in G. It is disjoint union of the sets

(4.3.7a) $G'_L = \bigcup_{g \in G} \mathrm{ad}(g)(L \cap G')$, L a Cartan subgroup of G,

where L ranges over a set of representatives of the conjugacy classes of Cartan subgroups. We denote

(4.3.7b) $(MA)'' = M''A$ regular set in MA.

Notice $MA \cap G' \subset (MA)''$. If J is a Cartan subgroup of MA, we denote

(4.3.7c) $(MA)''_J = M''_{J_M}A = \bigcup_{m \in M} (\mathrm{ad}(m)(J \cap M''))A$.

The following theorem unifies and extends various results of Bruhat [5, Chapter III], Harish-Chandra ([25, p. 544] and [26, §11]), Hirai [27, Theorems 1 and 2] and Lipsman [30, Theorem 9.1]. We assume the adjustment of Lemma 4.3.6. The specialization to the H-series is in §4.4.

<u>4.3.8. Theorem</u>. *Let* $\zeta \in \hat{Z}$, $[\eta] \in \hat{M}_\zeta$ *and* $\sigma \in \mathfrak{a}^*$. *Suppose that*

χ_ν *is the infinitesimal character of* $[\eta]$ *relative to* $\mathfrak{t}$. *Let* Ψ_η *denote the distribution character of* $[\eta]$.

1. $[\pi_{\eta,\sigma}]$ *has infinitesimal character* $\chi_{\nu+i\sigma}$ *relative to* $\mathfrak{h}$.

2. $[\pi_{\eta,\sigma}]$ *is a finite sum of classes from* $\hat{G}_\zeta$. *In particular,* $[\pi_{\eta,\sigma}]$ *has distribution character* $\Theta_{\pi_{\eta,\sigma}}$ *which is a locally summable function analytic on the regular set* G'.

3. $\Theta_{\pi_{\eta,\sigma}}$ *has support in the closure of* $\bigcup G_J'$ *where* J *runs over a system of representatives of the* G-*conjugacy classes of Cartan subgroups of* MA.

4. *Given a Cartan subgroup* $J = J_M \times A \subset MA$ *of* G, *let* $N_G(J)$ *and* $N_{MA}(J)$ *denote its normalizers in* G *and* MA. *If* $h \in J \cap G'$, *then* $N_G(J)(h)$ *and* $N_{MA}(J)(h)$ *are finite and*

$$\Theta_{\pi_{\eta,\sigma}}(h) = \frac{1}{|\Delta_{G,J}(h)|} \sum_{w(h) \in N_G(J)(h)} \frac{|\Delta_{MA,J}(wh)|}{|N_{MA}(J)(wh)|} \Psi_\eta((wh)_M) e^{i\sigma}((wh)_A).$$

{*Note that the sum runs over the Weyl group* $W_{G,J}$ *in case* $h \in J^0$.} *In particular, if* $ta \in H \cap G'$ *with* $t \in T$ *and* $a \in A$, *then*

$$\Theta_{\pi_{\eta,\sigma}}(ta) = \left|\frac{\Delta_{M,T}(t)}{\Delta_{G,H}(ta)}\right| \sum_{N_G(H)(ta)} |N_M(T)(wt)|^{-1} \Psi_\eta(wt) e^{i\sigma}(wa).$$

Parts 3 and 4 of Theorem 4.3.8 say that the distribution $\Theta_{\pi_{\eta,\sigma}}$ is independent of choice of cuspidal parabolic P associated to H. As $[\pi_{\eta,\sigma}]$ is a finite sum from $\hat{G}$ by part 2, now it also is independent of choice of P. So Theorem 4.3.8 implies

4.3.9. Corollary. *The class* $[\pi_{\eta,\sigma}]$ *is independent of the choice of cuspidal parabolic subgroup* $P = MAN$ *of* G *associated to the Cartan subgroup*

$H = T \times A$ *of* G.

The proof of Theorem 4.3.8 is distributed through the rest of §4.3. It is based on the following transform $C_c^\infty(G) \to C_c^\infty(MA)$, which is a minor variation on the Harish-Chandra transform.

4.3.10. Proposition. *If* $b \in C_c^\infty(G)$ *we obtain* $b_P \in C_c^\infty(MA)$ *by*

(4.3.11a) $$b_P(ma) = e^{-\rho_{\mathfrak{a}}}(a) \int_{K/Z} \left\{ \int_N b(kmank^{-1})dn \right\} d(kZ).$$

Then $\pi_{\eta,\sigma}(b)$ *is of trace class and*

(4.3.11b) $$\text{trace } \pi_{\eta,\sigma}(b) = \int_{MA} b_P(ma)\Psi_\eta(m)e^{i\sigma}(a)dmda.$$

Proof. Let K_1 denote the image of a Borel section to $K \to K/Z$. If $f \in L_2(\mathbf{E}_{\eta,\sigma})$ now f is specified by $f|_{K_1}$ because $G = KP$ and $Z \subset P$.

Compute

$$[\pi_{\eta,\sigma}(b)f](k') = \int_G b(g)f(g^{-1}k')dg = \int_G b(k'g)f(g^{-1})dg$$

$$= \int_{K/K\cap M} d(kM) \int_{MAN} b(k'mank^{-1})f(k(man)^{-1})\det(ad(a)_{\mathfrak{n}})dmdadn$$

$$= \int_{K/K\cap M} d(kM) \int_{MAN} b(k'mank^{-1})f(k(man)^{-1})e^{-2\rho_{\mathfrak{a}}}(a)dmdadn$$

$$= \int_{K/K\cap M} \left\{ \int_{MAN} b(k'mank^{-1})e^{-\rho_{\mathfrak{a}}+i\sigma}(a)\eta(m)dmdadn \right\} \cdot f(k)d(kM).$$

Now define a linear transformation of E_η by

$$\Phi_b(k',k) = \int_{MAN} b(k'mank^{-1})e^{-\rho_{\mathfrak{a}}+i\sigma}(a)\eta(m)dmdnda.$$

If $m_1 \in M \cap K$ then $f(km_1) = \eta(m_1)^{-1}f(k)$ and $\Phi_b(k',km_1) = \Phi_b(k',k)\eta(m_1)$; so

$$\Phi_b(k',km_1)f(km_1) = \Phi_b(k',k)f(k) \quad \text{for } m_1 \in M \cap K.$$

Now we have a well defined expression

$$[\pi_{\eta,\sigma}(b)f](k') = \int_{K/K\cap M} \Phi_b(k',k)f(k)d(kM).$$

As $Z \subset K \cap M$ and $K_1 \simeq K/Z$ compact, we write this as

$$[\pi_{\eta,\sigma}(b)f](k') = \int_{K_1} \Phi_b(k',k)f(k)d(kZ).$$

This operator has trace given by

$$\text{trace } \pi_{\eta,\sigma}(b) = \int_{K_1} \text{trace } \Phi_b(k,k)d(kZ)$$

in the sense that $\pi_{\eta,\sigma}(b)$ is trace class and the equation holds provided that the integral is absolutely convergent. Now set

$$\phi_b(k,m) = \int_{NA} e^{-\rho_{\mathfrak{a}}+i\sigma}(a)b(kmank^{-1})dnda.$$

Then $\phi_b \in C_c^\infty((K/Z) \times M)$. Noting that every integral involved is absolutely convergent, we calculate

$$\begin{aligned}
\int_{K_1} \text{trace } \Phi_b(k,k)d(kZ) &= \int_{K_1} \left\{ \text{trace} \int_M \phi_b(k,m)\eta(m)dm \right\} d(kZ) \\
&= \int_{K_1} \left\{ \int_M \phi_b(k,m)\Psi_\eta(m)dm \right\} d(kZ) \\
&= \int_{K_1} \left\{ \int_M \left\{ \int_{NA} e^{-\rho_{\mathfrak{a}}+i\sigma}(a)b(kmank^{-1})dnda \right\} \Psi_\eta(m)dm \right\} d(kZ) \\
&= \int_M \left\{ \int_A b_P(ma)e^{i\sigma}(a)\Psi_\eta(m)da \right\} dm.
\end{aligned}$$

That establishes Proposition 4.3.10. *q.e.d.*

We calculate the infinitesimal character of $\pi_{\eta,\sigma}$. As before, $\mathfrak{G}$ is the universal enveloping algebra of $\mathfrak{g}_C$ and $\mathfrak{Z}$ is the center of $\mathfrak{G}$. Let $\mathfrak{Z}_{MA}$ denote the center of the universal enveloping algebra of $(\mathfrak{m}+\mathfrak{a})_C$. Recall the canonical isomorphism

$$\gamma_G: \mathfrak{Z} \to W(\mathfrak{g}_C,\mathfrak{h}_C)\text{-invariant polynomials on } \mathfrak{h}_C^*.$$

Similarly we have an isomorphism

$$\gamma_{MA}: \mathfrak{Z}_{MA} \to W(\mathfrak{m}_C+\mathfrak{a}_C,\mathfrak{h}_C)\text{-invariant polynomials on } \mathfrak{h}_C^*.$$

Harish-Chandra [21, §12] showed that $\mu_{MA} = \gamma_{MA}^{-1}\cdot\gamma_G: \mathfrak{Z} \to \mathfrak{Z}_{MA}$ has the property: $\mathfrak{Z}_{MA}$ has finite rank over its subalgebra $\mu_{MA}(\mathfrak{Z})$.

Proposition 4.3.10 says trace $\pi_{\eta,\sigma}(b) = \text{trace } (\eta \otimes e^{i\sigma})(b_P)$ for $b \in C_c^\infty(G)$. Thus Harish-Chandra's result [21, Lemma 52] says

$$\text{trace } \pi_{\eta,\sigma}(zb) = \text{trace } (\eta \otimes e^{i\sigma})(\mu_{MA}(z)b_P) \quad \text{for} \quad z \in \mathfrak{Z}.$$

The infinitesimal character is thus given by

$$\begin{aligned}\chi_{\pi_{\eta,\sigma}}(z) &= \chi_{\eta\otimes e^{i\sigma}}(\mu_{MA}(z)) \\ &= \chi_{\nu+i\sigma}(\mu_{MA}(z)) \quad \text{for} \quad ((\mathfrak{m}+\mathfrak{a})_C, \mathfrak{h}_C). \\ &= \gamma_{MA}(\mu_{MA}(z))(\nu+i\sigma) = \gamma_G(z)(\nu+i\sigma) \\ &= \chi_{\nu+i\sigma}(z) \quad \text{for} \quad (\mathfrak{g}_C, \mathfrak{h}_C).\end{aligned}$$

That proves assertion 1 of Theorem 4.3.8.

We check that $[\pi_{\eta,\sigma}]$ is a direct integral over $\hat{G}_\zeta$. Adopt temporary notation $G^1 = ZG^0$ and $M^1 = M \cap G^1$. Since $[\eta] \in \hat{M}_\zeta$ we have $[\eta^1] \in (M^1)^\wedge_\zeta$ such that $[\eta]$ is a subrepresentation of $[\text{Ind}_{M^1\uparrow M}(\eta^1)]$. So $[\pi_{\eta,\sigma}]$ is a subrepresentation of

$$\text{Ind}_{P\uparrow G}\{(\text{Ind}_{M^1\uparrow M}(\eta^1)) \otimes e^{i\sigma}\} = \text{Ind}_{G^1\uparrow G}\{\text{Ind}_{M^1AN\uparrow G^1}(\eta^1 \otimes e^{i\sigma})\}.$$

However $[\eta^1] \in (M^1)^\wedge_\zeta$ implies $\text{Ind}_{M^1AN\uparrow G^1}(\eta^1 \otimes e^{i\sigma})|_Z$ to be a multiple of ζ. Thus $[\pi_{\eta,\sigma}]$ is a sum over $\hat{G}_\zeta$.

Now we are going to prove that there is an integer $n > 0$ such that

(4.3.12) $\pi_{\eta,\sigma}|_K = \sum_{\hat{K}_\zeta} m_\kappa \kappa$ where $0 \leqslant m_\kappa \leqslant n\cdot \dim \kappa < \infty$.

As in the discussion of (3.2.2), it suffices to prove this for the finite index subgroup ZG^0 of G. In other words we may suppose Z central in G for the proof of (4.3.12). Now (4.3.3) and the paragraph above give us a K-equivariant injective isometry

$$r_K\colon L_2(\boldsymbol{\mathcal{E}}_{\eta,\sigma}) \to L_2(K/Z,\zeta) \otimes E_\eta \quad \text{by} \quad r_K(f) = f|_K.$$

The result $L_2(K/Z,\zeta) = \sum_{\hat{K}_\zeta} V_\kappa \otimes V_\kappa^*$ of Lemma 2.4.2 allows us to identify the image. Using superscripts to denote fixed point sets now

$$r_K\colon L_2(\mathcal{E}_{\eta,\sigma}) \to \sum_{\hat{K}_\zeta} V_\kappa \otimes (V_\kappa^* \otimes E_\eta)^{M\cap K}$$

is a K-equivariant bijective isometry. Applying (3.2.2) to M and η we have

$$\eta|_{M\cap K} = \sum_{(M\cap K)^\wedge_\zeta} m_i \mu_i \quad \text{where } 0 \leq m_i \leq n_M \dim \mu_i < \infty.$$

If $[\kappa] \in \hat{K}_\zeta$, finite dimensionality gives us a finite sum

$$\kappa|_{M\cap K} = \sum_{(M\cap K)^\wedge_\zeta} m_{\kappa,i}\mu_i .$$

Now we have

$$\dim(V_\kappa^* \otimes E_\eta)^{M\cap K} = \sum m_{\kappa,i} m_i \leq n_M \sum m_{\kappa,i} \dim \mu_i = n_M \dim \kappa .$$

That proves (4.3.12). From the proof we note $n \leq n_M |G/ZG^0|$.

We now use an argument of Harish-Chandra to prove that $[\pi_{\eta,\sigma}]$ is a finite sum of classes from $\hat{G}_\zeta$. As in the discussion of (3.2.2) it suffices to consider the case where G is connected. Then Z is central, so $[\pi_{\eta,\sigma}]$ has central character ζ. We saw that $[\pi_{\eta,\sigma}]$ has infinitesimal character $\chi_{\nu+i\sigma}$ relative to $\mathfrak{h}$. Further (4.3.12) shows that $\pi_{\eta,\sigma}$ has no nontrivial subrepresentation of infinite multiplicity. Thus $\pi_{\eta,\sigma}$ is quasi-simple in the sense [20, p. 145] of Harish-Chandra. It follows that $[\pi_{\eta,\sigma}]$ has distribution character $\Theta_{\pi_{\eta,\sigma}}$ that is a locally summable function analytic on the regular set G' [20, Theorem 6], and that $\pi_{\eta,\sigma} = \sum \pi_j$ discrete direct sum of irreducible representations [16, Lemma 2]. Notice $[\pi_j] \in \hat{G}_\zeta$ and $\chi_{\pi_j} = \chi_{\pi_{\eta,\sigma}} = \chi_{\nu+i\sigma}$. Harish-Chandra proved [20, Theorem 7] that the differential equations

$$z\Theta = \chi_{\nu+i\sigma}(z)\Theta \quad \text{for all } z \in \mathfrak{Z}$$

constrain the Θ_{π_j} to a finite dimensional space of functions on G. However, inequivalent classes in $\hat{G}_\zeta$ have linearly independent distribution characters. Now $\pi_{\eta,\sigma} = \sum \pi_j$ involves only finitely many classes from $\hat{G}_\zeta$.

But (4.3.1) forces the multiplicities $m(\pi_j,\pi) < \infty$. Now $\pi_{\eta,\sigma}$ is a finite sum from $\hat{G}_\zeta$. We have proved assertion 2 of Theorem 4.3.8.

Now we start to calculate $\Theta_{\pi_{\eta,\sigma}}$ by adapting Lipsman's argument [30, Theorem 9.1] to our more general situation. Recall our notation (4.3.7) and (4.3.11a).

<u>4.3.13. Lemma</u>. *If* $b \in C_c^\infty(G_L')$ *where* L *is a Cartan subgroup of* G *that is not conjugate to a subgroup of* MA, *then* $b_P = 0$. *If* $b \in C_c^\infty(G_J')$ *where* J *is a Cartan subgroup of* MA, *then* $b_P \in C_c^\infty(MA \cap G_J') \subset C_c^\infty((MA)_J'')$.

Proof. If $ma \in MA$ and $ad(ma)^{-1} - 1$ is nonsingular on $\mathfrak{n}$, then [22, Lemma 11] and (3.1.1) give

$$\int_N b(kmank^{-1})dn = |\det(ad(ma)^{-1}-1)_{\mathfrak{n}}| \int_N b(knman^{-1}k^{-1})dn.$$

If $b \in C_c^\infty(G_L')$ where G_L' does not meet MA, then $b(knman^{-1}k^{-1})$ is identically zero, so $b_P = 0$. If $b \in C_c^\infty(G_J')$ where $J \subset MA$, then there is a compact set $S \subset G_J'$ such that

$$\text{if } \int_N b(kmank^{-1})dn \neq 0 \text{ for some } k \in K, \text{ then } ma \in S.$$

Thus b_P is supported in $S \cap MA \subset G_J' \cap MA \subset (MA)_J''$. *q.e.d.*

Let L be a Cartan subgroup of G not conjugate to a Cartan subgroup of MA. If $b \in C_c^\infty(G_L')$, we combine Proposition 4.3.10 and Lemma 4.3.13 to see $\Theta_{\pi_{\eta,\sigma}}(b) = \int_{MA} b_P(ma)\Psi_\eta(m)e^{i\sigma}(a)dmda = 0$. Thus $\Theta_{\pi_{\eta,\sigma}}|_{G_L'} = 0$. That proves assertion 3 of Theorem 4.3.8.

Fix a Cartan subgroup $J = J_M \times A \subset MA$ of G. To calculate $\Theta_{\pi_{\eta,\sigma}}$ on G_J' we need a variation on the Weyl Integration Formula.

Retain the notation (4.3.4) and (4.3.7). Let Z_J denote the center of J. It is open in J, and thus inherits Haar measure from the Haar measure dh on J. Normalize the G-invariant measure on G/Z_J by

$$\int_G f(g)dg = \int_{G/Z_J}\left\{\int_{Z_J} f(gh)dh\right\}d(gZ_J) \quad \text{for} \quad f \in C_c(G).$$

It turns out that the argument of Weyl extends, as follows, even farther than remarked by Harish-Chandra [24, Lemma 91].

4.3.14. Lemma. *If* $b \in C_c(G_J')$ *then*

$$\int_G b(g)dg = \int_{J\cap G'} |N_G(J)(h)|^{-1}\left\{\int_{G/Z_J} b(ghg^{-1})d(gZ_J)\right\}|\Delta_{G,J}(h)|^2 dh.$$

Proof. The map $(G/Z_J) \times (J\cap G') \to G_J'$, given by $(gZ_J,h) \to ghg^{-1}$, is regular and surjective, and is $|N_G(J)(h)|$ to one on ghg^{-1}. Its Jacobian determinant at (gZ_H,h) is

$$|\det(\mathrm{ad}(h)-1)_{\mathfrak{g}/\mathfrak{j}}| = \left|\prod_{\gamma\in\Sigma_j}(e^{\gamma}-1)(h)\right|, \quad \Sigma = \Sigma_j^+ \cup -\Sigma_j^+.$$

However

$$\prod_{\gamma\in\Sigma_j}(e^{\gamma}-1) = \prod_{\gamma\in\Sigma_j^+}(e^{\gamma}-1)\cdot\prod_{\gamma\in\Sigma_j^+}(e^{-\gamma}-1)$$

$$= \left\{e^{\rho_j}\prod_{\gamma\in\Sigma_j^+}(e^{\gamma/2}-e^{-\gamma/2})\right\}\left\{e^{-\rho_j}\prod_{\gamma\in\Sigma_j^+}(e^{-\gamma/2}-e^{\gamma/2})\right\}$$

$$= (-1)^n(\Delta_{G,J})^2 \quad \text{where} \quad n = |\Sigma_j^+| = \tfrac{1}{2}\dim(\mathfrak{g}/\mathfrak{j}).$$

Now that Jacobian determinant is $|\Delta_{G,J}(h)|^2$. *q.e.d.*

Similarly, normalize invariant measure on MA/Z_J by the condition

$$\int_{MA} F(ma)dmda = \int_{MA/Z_J}\left\{\int_{Z_J} F(mah)dh\right\}d(maZ_J) \quad \text{for} \quad F \in C_c(MA),$$

and then

4.3.15. Lemma. *If* $B \in C_c((MA)_J'')$ *then*

$$\int_{MA} B(ma)dmda = \int_{J\cap MA''} |N_{MA}(J)(h)|^{-1}\left\{\int_{MA/Z_J} (maha^{-1}m^{-1})d(maZ_J)\right\}|\Delta_{MA,J}(h)|^2 dh.$$

Fix $b \in C_C^\infty(G_J')$. Proposition 4.3.10 and existence of $\Theta_{\pi_{\eta,\sigma}}$ give us

$$\Theta_{\pi_{\eta,\sigma}}(b) = \int_{MA} b_P(ma)\Psi_\eta(m)e^{i\sigma}(a)\,dm\,da.$$

Lemma 4.3.13 says $b_P \in C_c^\infty(MA \cap G_J') \subset C_c^\infty((MA)_J'')$. Since $ma \to \Psi_\eta(m)e^{i\sigma}(a)$ is a class function on MA, Lemma 4.3.15 now expresses $\Theta_{\pi_{\eta,\sigma}}(b)$ as

(4.3.16a)

$$\int_{J\cap G'} |N_{MA}(J)(h)|^{-1}\left\{\int_{MA/Z_J} b_P(maha^{-1}m^{-1})d(maZ_J)\right\}\Psi_\eta(h_M)e^{i\sigma}(h_A)|\Delta_{MA,J}(h)|^2dh$$

where $h = h_M h_A$ according to $J = J_M \times A$. Since A is central in MA, the definition (4.3.11a) of b_P forces

(4.3.16b)

$$\int_{MA/Z_J} b_P(maha^{-1}m^{-1})d(maZ_J)$$

$$= e^{-\rho_{\mathfrak{a}}}(h_A)\int_{MA/Z_J} d(maZ_J)\cdot\int_{K/Z} d(kZ)\cdot\int_N b(k\cdot maha^{-1}m^{-1}\cdot n\cdot k^{-1})dn.$$

Unimodularity of N and [22, Corollary to Lemma 11] say

(4.3.16c)

$$\int_N b(k\cdot maha^{-1}m^{-1}\cdot n\cdot k^{-1})dn$$

$$= |\det(\mathrm{ad}(h^{-1})-1)_{\mathfrak{n}}|\int_N b(knma\cdot h\cdot(knma)^{-1})dn.$$

We modify Harish-Chandra's evaluation [22, Lemma 12] of $|\det(\mathrm{ad}(h^{-1})-1)_{\mathfrak{n}}|$ for the case $J = H$. Choose $c \in \mathrm{Int}(\mathfrak{g}_C)$ that sends $\mathfrak{j}_C$ onto $\mathfrak{h}_C$, such that $c(x) = x$ for $x \in \mathfrak{a}$ and $c^*\Sigma^+ = \Sigma_{\mathfrak{j}}^+$. Then $c(\mathfrak{m}_C) = \mathfrak{m}_C$ and $c(\mathfrak{n}_C) = \mathfrak{n}_C$, and we compute

$$\det(\mathrm{ad}(h^{-1})-1)_{\mathfrak{n}} = \det(\mathrm{ad}(c(h^{-1})-1)_{\mathfrak{n}}$$

$$= \prod_{\gamma\in\Sigma^+-\Sigma_{\mathfrak{t}}^+}(e^{-\gamma}(c(h^{-1}))-1) = \prod_{\beta\in\Sigma_{\mathfrak{j}}^+-\Sigma_{\mathfrak{j}_M}^+}(e^{\beta}(h)-1)$$

$$= (e^{\rho_{\mathfrak{j}}}(h)\Delta_{G,J}(h))/(e^{\rho_{\mathfrak{j}_M}}(h)\Delta_{MA,J}(h)).$$

On the other hand, $e^{\rho_j}(h)/e^{\rho_{j_M}}(h)$ is the product of $e^{\rho_{\mathfrak{a}}}(h_A)$ with $(\prod_{\beta\in\Sigma_j^+-\Sigma_{j_M}^+} e^{\beta/2})(h_M) = \{\det \mathrm{ad}(h_M)_{\mathfrak{n}}\}^{-1/2}$. Since $\mathrm{ad}(M)$ acts on $\mathfrak{n}$ with determinant of absolute value 1, now $|e^{\rho_j}(h)/e^{\rho_{j_M}}(h)| = |e^{\rho_{\mathfrak{a}}}(h_A)| = e^{\rho_{\mathfrak{a}}}(h_A)$. In summary,

(4.3.17) $$|\det(\mathrm{ad}(h^{-1})-1)_{\mathfrak{n}}| = e^{\rho_{\mathfrak{a}}}(h_A)\,|\Delta_{G,J}(h)/\Delta_{MA,J}(h)|.$$

Substitute (4.3.17) into (4.3.16c) and then (4.3.16b). The result is

$$\begin{aligned}
&\int_{MA/Z_J} b_P(maha^{-1}m^{-1})\,d(maZ_J)\\
&= \int_{MA/Z_J} d(maZ_J)\cdot\int_{K/Z} d(kZ)\cdot|\Delta_{G,J}(h)/\Delta_{MA,J}(h)|\int_N b(\mathrm{ad}(knma)h)\,dn\\
&= |\Delta_{G,J}(h)/\Delta_{MA,J}(h)|\int_{K/K\cap M}\left\{\int_{MNA/Z_J} b(\mathrm{ad}(kmna)h)\,dm\,da\,dn\right\} d(kM)\\
&= |\Delta_{G,J}(h)/\Delta_{MA,J}(h)|\int_{G/Z_J} b(ghg^{-1})\,d(gZ_J).
\end{aligned}$$

Substituting into (4.3.16a), now

(4.3.18) $$\Theta_{\pi_{\eta,\sigma}}(b) = \int_{J\cap G'} |N_{MA}(J)(h)|^{-1}\left\{\int_{G/Z_J} b(ghg^{-1})\,d(gZ_J)\right\}\Psi_\eta(h_M)e^{i\sigma}(h_A)\,|\Delta_{G,J}(h)\Delta_{MA,J}(h)|\,dh.$$

Define an $\mathrm{ad}(N_G(J))$-invariant function on $J\cap G'$ by

(4.3.19) $$\Phi_{\eta,\sigma,J}(h) = \frac{1}{|\Delta_{G,J}(h)|}\sum_{w(h)\in N_G(J)(h)} \frac{|\Delta_{MA,J}(wh)|}{|N_{MA}(J)(wh)|}\,\Psi_\eta((wh)_M)\,e^{i\sigma}((wh)_A).$$

This extends uniquely to a class function (same notation) on G_J'. Substituting that class function into (4.3.18),

$$\Theta_{\pi_{\eta,\sigma}}(b)$$

$$= \int_{J\cap G'} |N_G(J)(h)|^{-1}\left\{\int_{G/Z_J} b(ghg^{-1})\Phi_{\eta,\sigma,J}(ghg^{-1})d(gZ_J)\right\}|\Delta_{G,J}(h)|^2 dh.$$

Now Lemma 4.3.14 says $\Theta_{\pi_{\eta,\sigma}}(b) = \int_G b(g)\Phi_{\eta,\sigma,J}(g)dg$. As b was an arbitrary element of $C_c^\infty(G_J')$, we conclude that $\Theta_{\pi_{\eta,\sigma}}|_{G_J'}$ is given by the formula (4.3.19) for $\Phi_{\eta,\sigma,J}$. That proves the general formula of assertion 4 of Theorem 4.3.8.

Suppose $J = H$. Then $J_M = T$ and A are separately normalized by $N_G(J) = N_G(H)$. Writing $t = h_M$ and $a = h_A$ now each $|\Delta_{MA,H}(wh)| = |\Delta_{MA,H}(h)| = |\Delta_{M,T}(t)|$ and $(wh)_M = w(t)$ and $(wh)_A = w(a)$. Finally, since A is central in MA, it is pointwise fixed under $N_{MA}(H)$, so $N_{MA}(H) = N_M(T)A$ and $|N_{MA}(H)(wh)| = |N_M(T)(wt)|$. Substituting into the general formula, this completes the proof of assertion 4.

Theorem 4.3.8 is proved. *q.e.d.*

4.4. We specialize the results of §4.3 to the H-series of G by means of our rather precise knowledge of $\hat{M}_{disc}$.

The Cartan subgroup $H = T \times A$ and the associated cuspidal parabolic subgroup $P = MAN$ are fixed. That determines the positive $\mathfrak{a}$-root system $\Sigma_{\mathfrak{a}}^+$ on $\mathfrak{g}$ for which $\mathfrak{n}$ is the sum of the negative $\mathfrak{a}$-root spaces. Choose a positive $\mathfrak{t}_C$-root system $\Sigma_{\mathfrak{t}}^+$ on $\mathfrak{m}_C$. Let Σ^+ be the positive $\mathfrak{h}_C$-root system on $\mathfrak{g}_C$ specified by Lemma 4.1.7 that satisfies (4.3.4a). Define ρ, $\rho_{\mathfrak{a}}$, $\rho_{\mathfrak{t}}$, $\Delta_{G,H}$ and $\Delta_{M,T}$ as in (4.3.4). Make the adjustment of Lemma 4.3.6, if necessary. Thus $e^\rho \in \hat{H}$, $e^{\rho_{\mathfrak{t}}} \in \hat{T}$ and $e^{\rho_{\mathfrak{t}}}(Z) = e^\rho(Z) = 1$ so that $\Delta_{G,H}$ is well defined on H and $\Delta_{M,T}$ is well defined on T.

The description of $\hat{G}_{disc}$ in §§3.4 and 3.5 is valid for $\hat{M}_{disc}$, according to Proposition 4.1.6. It comes out as follows. Define

$$(4.4.1a) \quad \tilde{\omega}_{\mathfrak{t}}(\nu) = \prod_{\phi\in\Sigma_{\mathfrak{t}}^+} \langle \phi,\nu\rangle \quad \text{for } \nu \in \mathfrak{t}_C^*$$

and

$$(4.4.1b) \quad L_{\mathfrak{t}}'' = \{\nu \in i\mathfrak{t}^* : e^{\nu} \text{ is defined on } T^0 \text{ and } \tilde{\omega}_{\mathfrak{t}}(\nu) \neq 0\}.$$

If $\nu \in L_{\mathfrak{t}}''$ there is a unique class $[\eta_\nu] \in (M^0)^\wedge_{disc}$ whose distribution character (say Ψ_{η_ν}) satisfies

$$(4.4.2) \quad \Psi_{\eta_\nu}\big|_{T^0\cap M''} = (-1)^{q_M(\nu)} \frac{1}{\Delta_{M,T}} \sum_{w\in W_{M^0,T^0}} \det(w)e^{w\nu}$$

where q_M is defined on $L_{\mathfrak{t}}''$ in the manner of (3.4.6) using $\Sigma_{\mathfrak{t}}^+$, and where M'' denotes the M-regular subset of M. Every class in $(M^0)^\wedge_{disc}$ is one of the $[\eta_\nu]$. Classes $[\eta_\nu] = [\eta_{\nu'}]$ precisely when $\nu' \in W_{M^0,T^0}(\nu)$. Finally $[\eta_\nu]$ has central character $e^{\nu-\rho_{\mathfrak{t}}}|_{Z_{M^0}}$ and has infinitesimal character χ_ν relative to $\mathfrak{t}$. Compare with Theorem 3.4.7 and Corollary 3.4.9.

If $\nu \in L_{\mathfrak{t}}''$ and $[\chi] \in Z_M(M^0)^\wedge_\xi$ where $\xi = e^{\nu-\rho_{\mathfrak{t}}}|_{Z_{M^0}}$, then we define

$$(4.4.3a) \quad \eta_{\chi,\nu} = \operatorname{Ind}_{M^\dagger\uparrow M}(\chi \otimes \eta_\nu)$$

where $[\eta_\nu] \in (M^0)^\wedge_{disc}$ as in (4.4.2). Then $[\eta_{\chi,\nu}] \in \hat{M}_{disc}$, and it is the unique class there whose distribution character $\Psi_{\eta_{\chi,\nu}}$ is given on $Z_M(M^0)\cdot(T^0\cap M'')$ by

$$(4.4.3b) \quad \Psi_{\eta_{\chi,\nu}}(zt) =$$

$$= \sum_{1\leqslant j\leqslant r} (-1)^{q_M(w_j\nu)} \operatorname{tr} \chi(x_j^{-1}zx_j)\cdot\frac{1}{\Delta_{M,T}} \sum_{W_{M^0,T^0}} \det(ww_j)e^{ww_j\nu}(t).$$

Here the $w_j = \mathrm{ad}(x_j T)_{\mathfrak{t}}$ are representatives of $W_{M,T}$ modulo W_{M^0,T^0}. Every class in $\hat{M}_{\mathrm{disc}}$ is one of the $[\eta_{\chi,\nu}]$. Classes $[\eta_{\chi,\nu}] = [\eta_{\chi',\nu'}]$ precisely when $([\chi'],\nu) \in W_{M,T}([\chi],\nu)$. Finally $[\eta_{\chi,\nu}]$ has infinitesimal character χ_ν relative to $\mathfrak{t}$.

We combine this description with Theorem 4.3.8. Recall that the normalizers $N_{MA}(H) = N_M(T) \times A$ and $N_G(H)$ have all orbits finite on $H \cap G'$.

<u>4.4.4 Theorem</u>. *Let* $\nu \in L''_{\mathfrak{t}}$ *and* $[\chi] \in Z_M(M^0)^{\wedge}_{\xi}$ *where* $\xi = e^{\nu-\rho_{\mathfrak{t}}}|_{Z_{M^0}}$. *Let* $\sigma \in \mathfrak{a}^*$. *Define* $[\eta_{\chi,\nu}] \in \hat{M}_{\mathrm{disc}}$ *and its character* $\Psi_{\eta_{\chi,\nu}}$ *by* (4.4.3). *Then* $[\pi_{\chi,\nu,\sigma}] = [\mathrm{Ind}_{P\uparrow G}(\eta_{\chi,\nu} \otimes e^{i\sigma})]$ *is the unique* H-*series representation class on* G *whose distribution character satisfies*

$$\Theta_{\pi_{\chi,\nu,\sigma}}(ta) = \left|\frac{\Delta_{M,T}(t)}{\Delta_{G,H}(ta)}\right| \sum_{w(ta)\in N_G(H)(ta)} |N_M(T)(wt)|^{-1}\Psi_{\eta_{\chi,\nu}}(wt)e^{i\sigma}(wa)$$

for $ta \in H\cap G'$ *with* $t \in T$ *and* $a \in A$.

Every H-*series class on* G *is one of the* $[\pi_{\chi,\nu,\sigma}]$. *Classes* $[\pi_{\chi,\nu,\sigma}] = [\pi_{\chi',\nu',\sigma'}]$ *precisely when* $([\chi'],\nu',\sigma')$ *is in the Weyl group orbit* $W_{G,H}([\chi],\nu,\sigma)$.

$[\pi_{\chi,\nu,\sigma}]$ *is a finite sum from* $\hat{G}_\zeta$ *where* $[\eta_{\chi,\nu}] \in \hat{M}_\zeta$. *Dual class* $[\pi_{\chi,\nu,\sigma}{}^*] = [\pi_{\bar{\chi},-\nu,-\sigma}]$. *The class* $[\pi_{\chi,\nu,\sigma}]$ *has infinitesimal character* $\chi_{\nu+i\sigma}$ *relative to* $\mathfrak{h}$; *so it sends the Casimir element of* $\mathfrak{g}$ *to* $\|\nu\|^2 + \|\sigma\|^2 - \|\rho\|^2$.

<u>4.4.5 Corollary</u>. *Each* H-*series class* $[\pi_{\chi,\nu,\sigma}]$, *and thus also the* H-*series of* G, *is independent of choice of cuspidal parabolic subgroup* P *associated to* H.

Proof. The only assertion that does not follow directly from (4.4.3) and Theorem 4.3.8, is the assertion that $\Theta_{\pi_{\chi,\nu,\sigma}}\big|_{H\cap G'}$ determines $([\chi],\nu,\sigma)$ modulo $W_{G,H}$, and thus determines $[\pi_{\chi,\nu,\sigma}]$.

Suppose that $[\pi_{\chi,\nu,\sigma}]$ and $[\pi_{\chi',\nu',\sigma'}]$ are H-series classes whose characters have the same restriction to $H \cap G'$. Then

$$\sum |N_M(T)(wt)|^{-1}\psi_{\eta_{\chi,\nu}}(wt)e^{i\sigma}(wa) = \sum |N_M(T)(wt)|^{-1}\psi_{\eta_{\chi',\nu'}}(wt)e^{i\sigma'}(wa)$$

for $ta \in H \cap G'$, where the sum runs over $w(ta) \in N_G(H)(ta)$. By linear independence of the characters $e^{i\sigma''}$ on A, $\sigma'' \in \mathfrak{a}^*$, we may replace σ' by an element of $N_G(H)(\sigma') = W_{G,H}(\sigma')$ and suppose $\sigma' = \sigma$. Thus, on $H \cap G'$,

$$\sum |N_M(T)(wt)|^{-1}e^{i\sigma}(wa)\left\{\psi_{\eta_{\chi,\nu}}(wt) - \psi_{\eta_{\chi',\nu'}}(wt)\right\} = 0.$$

The functions $\psi_{\eta_{\chi'',\nu''}}$ on $T \cap M''$, $[\eta_{\chi'',\nu''}] \in \hat{M}_{disc}$, are linearly independent. Also $|N_M(T)(t_1)|$ is constant as t_1 ranges over $t_1T^0 \cap M''$, thus is locally constant on $T \cap M''$. We conclude that $\psi_{\eta_{\chi,\nu}} = \psi_{\eta_{\chi',\nu'}}$, which implies $([\chi'],\nu') \in W_{M,T}([\chi],\nu)$. *q.e.d.*

Notice that the support of $\Theta_{\pi_{\chi,\nu,\sigma}}$ meets the interior of G_H'. In view of Theorem 4.3.8(3), it determines the conjugacy class of H. There is a much stronger result, due to Lipsman [30, Theorem 11.1] for connected semisimple groups of finite center, which goes over to our case as follows.

4.4.6. Theorem. *Let* H *and* 'H *be Cartan subgroups of* G *that are not conjugate. Let* $[\pi]$ *be an* H-*series class and let* $['\pi]$ *be a* 'H-*series class.*

1. *Infinitesimal characters* $\chi_\pi \neq \chi_{'\pi}$.

2. $[\pi]$ *and* $['\pi]$ *are disjoint,* i.e. *do not have any nontrivial equivalent subrepresentations.*

Proof. We may take both H and 'H stable under the Cartan involution

θ. Let $H = T \times A$ and $'H = 'T \times 'A$ be the resultant splittings. Express $[\pi] = [\pi_{\chi,\nu,\sigma}]$ using H and $['\pi] = [\pi_{'\chi,'\nu,'\sigma}]$ using $'H$. Now the infinitesimal characters $\chi_\pi = \chi_{\nu+i\sigma}$ rel $\mathfrak{h}$ and $\chi_{'\pi} = \chi_{'\nu+i\cdot'\sigma}$ rel $'\mathfrak{h}$.

Suppose $\chi_\pi = \chi_{'\pi}$. Then $\mathfrak{g}_C$ has an inner automorphism β such that $\beta(\mathfrak{h}_C) = {'\mathfrak{h}}_C$ and $\beta^*('\nu + i\cdot'\sigma) = \nu + i\sigma$. But β is trivial on the center of $\mathfrak{g}_C$, and β^* carries the real span of the $'\mathfrak{h}_C$-roots to the real span of the $\mathfrak{h}_C$-roots. Thus $\beta(i\mathfrak{t} + \mathfrak{a}) = i\cdot'\mathfrak{t} + '\mathfrak{a}$. Now $\beta^*('\nu) = \nu$.

Choose a positive $\mathfrak{h}_C$-root system Σ^+ on $\mathfrak{g}_C$ such that $\Sigma^+_{\mathfrak{a}} = \{\gamma|_{\mathfrak{a}}: \gamma \in \Sigma^+$ and $\gamma|_{\mathfrak{a}} \neq 0\}$ is a positive $\mathfrak{a}$-root system on $\mathfrak{g}$ and $\Sigma^+_{\mathfrak{t}} = \{\gamma|_{\mathfrak{t}}: \gamma \in \Sigma^+$ and $\gamma|_{\mathfrak{a}} = 0\}$ is the positive $\mathfrak{t}_C$-root system on $\mathfrak{m}_C$ consisting of all $\mathfrak{t}_C$-roots ψ with $\langle\psi,\nu\rangle > 0$. Similarly choose $'\Sigma^+$ for $'\mathfrak{h}_C$ and $'\nu$. Either $\beta^*('\Sigma^+) = \Sigma^+$ or $\beta^*('\Sigma^+) \neq \Sigma^+$. In the latter case, since $\beta^*('\nu) = \nu$, there exists $w^* \in W(\mathfrak{g}_C, \mathfrak{h}_C)$ which is a product of reflections in root hyperplanes $\gamma^\perp$ containing ν, such that $w^*(\beta^*('\Sigma^+)) = \Sigma^+$, and we replace β by $\beta\cdot w$. Now we may (and do) assume $\beta^*('\Sigma^+) = \Sigma^+$.

Let Π be the simple root system for Σ^+ and express $\nu = \sum_{\gamma\in\Pi} n_\gamma(\nu)\cdot\gamma$. Then $n_\gamma(\nu) \neq 0$ precisely when $\gamma \in \Pi \cap \Sigma^+_{\mathfrak{t}}$. Let $'\Pi$ be the simple root system for $'\Sigma^+$ and make the analogous observations. Now

$$\beta(\mathfrak{a}_C) = \beta((\Pi \cap \Sigma^+_{\mathfrak{t}})^\perp) = ('\Pi \cap '\Sigma^+_{'\mathfrak{t}})^\perp = '\mathfrak{a}_C.$$

Since $\beta(\mathfrak{t} + i\mathfrak{a}) = '\mathfrak{t} + i\cdot'\mathfrak{a}$ we conclude $\beta(\mathfrak{a}) = '\mathfrak{a}$. Again $\beta(\mathfrak{t}_C) = \beta(\mathfrak{a}^\perp) = '\mathfrak{a}^\perp = '\mathfrak{t}_C$ so $\beta(\mathfrak{t}) = '\mathfrak{t}$. Now $\beta(\mathfrak{h}) = '\mathfrak{h}$.

We have shown that the inner automorphism β of $\mathfrak{g}_C$ sends $\mathfrak{h}$ to $'\mathfrak{h}$. Consequently [42, Corollary 2.4] there is an inner automorphism of $\mathfrak{g}$ that sends $\mathfrak{h}$ to $'\mathfrak{h}$. That contradicts nonconjugacy of H and $'H$. In summary, $\chi_\pi = \chi_{'\pi}$ is impossible.

Assertion 1 is proved. Assertion 2 follows because any nontrivial subrepresentation has the same infinitesimal character. *q.e.d.*

4.5. We complete our description of the H-series with the statement of an irreducibility theorem of Harish-Chandra. This theorem appears in an unpublished manuscript in which Harish-Chandra extends the method used by Bruhat [5,§7] for the principal and complementary series.

Fix a Cartan subgroup $H = T \times A$ and an associated cuspidal parabolic subgroup $P = MAN$ of G. The latter specifies the positive $\mathfrak{a}$-root system $\Sigma_{\mathfrak{a}}^+$ on $\mathfrak{g}$ such that $\mathfrak{n}$ is the sum of the negative $\mathfrak{a}$-root spaces. Choose a positive $\mathfrak{t}_C$-root system on $\mathfrak{m}_C$. Let Σ^+ be the positive $\mathfrak{h}_C$-root system on $\mathfrak{g}_C$ specified by Lemma 4.1.7.

Let $[\eta] \in \hat{M}$ have infinitesimal character χ_ν relative to $\mathfrak{t}_C$. If $\langle \phi, \nu \rangle$ is real for every root $\phi \in \Sigma_{\mathfrak{t}}^+$, then one says that $[\eta]$ has <u>real infinitesimal character</u>. For example, the classes in $\hat{M}_{disc}$ have real infinitesimal character.

An element $\sigma \in \mathfrak{a}^*$ is $(\mathfrak{g},\mathfrak{a})$-<u>regular</u> if $\langle \psi, \sigma \rangle \neq 0$ for every $\psi \in \Sigma_{\mathfrak{a}}^+$. To formulate some equivalent conditions, choose a minimal parabolic subgroup $P_0 = M_0A_0N_0$ of G with $A \subset A_0 = \theta A_0$. Recall the little Weyl group $W(\mathfrak{g}, \mathfrak{a}_0)$, normalizer-by-centralizer quotient $N_{G^0}(A_0)/Z_{G^0}(A_0)$; as group of linear transformations of $\mathfrak{a}_0^*$ it is generated by the reflections in the $\mathfrak{a}_0$-roots. The $\mathfrak{a}$-roots are just the nonzero restrictions of $\mathfrak{h}_C$-roots, and the corresponding result holds for $\mathfrak{a}_0$-roots; it follows that the $\mathfrak{a}$-roots are just the nonzero restrictions of $\mathfrak{a}_0$-roots. If $w \in W(\mathfrak{g}, \mathfrak{a}_0)$ and if $\sigma \in \mathfrak{a}^*$ is $(\mathfrak{g},\mathfrak{a})$-regular, then $\mathfrak{a}$ is central in the centralizer $\mathfrak{g}^\sigma$ and $w \in W(\mathfrak{g}^\sigma, \mathfrak{a}_0)$, so w is generated by reflections in roots that annihilate $\mathfrak{a}$, forcing $w|_{\mathfrak{a}}$ trivial. In summary:

<u>4.5.1. Lemma</u>. *If* $\sigma \in \mathfrak{a}^*$ *then the following conditions are*

equivalent.

(i) σ *is* $(\mathfrak{g}, \mathfrak{a})$-*regular.*

(ii) *If* $\phi \in \Sigma^+$ *and* $\phi|_{\mathfrak{a}} \neq 0$ *then* $\langle \phi, \sigma \rangle \neq 0$.

(iii) *If* ψ_0 *is an* $\mathfrak{a}_0$-*root of* $\mathfrak{g}$ *and* $\psi_0|_{\mathfrak{a}} \neq 0$ *then* $\langle \psi_0, \sigma \rangle \neq 0$.

(iv) *If* $w \in W(\mathfrak{g}, \mathfrak{a}_0)$ *and* $w|_{\mathfrak{a}}$ *is nontrivial then* $w(\sigma) \neq \sigma$.

The following irreducibility theorem is proved by Harish-Chandra (unpublished) in the generality of our case:

4.5.2. Theorem. *Let* $[\eta] \in \hat{M}$ *have real infinitesimal character and let* $\sigma \in \mathfrak{a}^*$ *be* $(\mathfrak{g}, \mathfrak{a})$-*regular. Then* $[\pi_{\eta,\sigma}] = [\mathrm{Ind}_{P\uparrow G}(\eta \otimes e^{i\sigma})]$ *is irreducible.*

An immediate consequence:

4.5.3. Corollary. *If* $\sigma \in \mathfrak{a}^*$ *is* $(\mathfrak{g}, \mathfrak{a})$-*regular, then every* H-*series class* $[\pi_{\chi,\nu,\sigma}]$ *is irreducible.*

§5. The Plancherel Formula for Reductive Lie Groups.

Let G be a reductive Lie group of the class described in §3.1. Fix a Cartan involution θ of G and a system $\{H_1,\dots,H_\ell\}$ of θ-stable representatives of the conjugacy classes of Cartan subgroups of G. Harish-Chandra proved a Plancherel formula [26, §11] for the case where the identity component G^0 has finite index in G and its derived group $[G^0,G^0]$ has finite center. In our notation (§4), Harish-Chandra's formula says that there are unique continuous functions $m_{j,\eta}$ on $\mathfrak{a}_j^*$, $[\eta] \in (M_j)^\wedge_{disc}$, invariant by the Weyl group W_{G,H_j}, such that

$$f(x) = \sum_{1\leq j\leq \ell} \sum_{[\eta]\in(M_j)^\wedge_{disc}} \deg(\eta) \int_{\mathfrak{a}_j^*} \Theta_{\pi_{\eta,\sigma}}(r_x f) m_{j,\eta}(\sigma)\, d\sigma$$

(absolutely convergent) for $f \in C_c^\infty(G)$. Further, the $m_{j,\eta}$ are restrictions of meromorphic functions on $\mathfrak{a}_{jC}^*$.

We extend a variation on Harish-Chandra's formula to the full class of groups described in §3.1. In fact one can drop (3.1.1). Our result is stronger than Harish-Chandra's in the sense that, given $\zeta \in \hat{Z}$, it applies to functions $f \in L_2(G/Z,\zeta)$ that are C^∞ with support compact modulo Z. Our result is weaker than Harish-Chandra's in the sense that the Plancherel densities $m_{j,\zeta,\nu}$ are only proved Borel measurable on $\mathfrak{a}_j^*$, as opposed to continuous on $\mathfrak{a}_j^*$ and meromorphic on $\mathfrak{a}_{jC}^*$. The result is formulated and stated in §5.1.

Our method is moderately delicate reduction to Harish-Chandra's Plancherel formula for G/Z. This is the content of §§5.2 through 5.6. In fact our proof should be considered provisional: if one carefully follows the details of Harish-Chandra's argument, he should be able to extend that argument to our case, proving the $m_{j,\zeta,\nu}$ meromorphic as well.

Finally, in §5.7, we extract some consequences of the Plancherel formula that play a key role in our geometric realizations of unitary representations.

5.1. We formulate the Plancherel Theorem for reductive Lie groups G that satisfy the conditions of §3.1.

Fix a Cartan involution θ of G; let K denote its fixed point set. Choose a complete system of θ-stable representatives

(5.1.1) $H_j = T_j \times A_j, \quad 1 \leqslant j \leqslant \ell,$

of the conjugacy classes of Cartan subgroups of G. Choices $\Sigma^+_{\mathfrak{a}_j}$ of positive $\mathfrak{a}_j$-root system on $\mathfrak{g}$ then specify representatives

(5.1.2) $P_j = M_jA_jN_j, \quad M_j \times A_j = Z_G(A_j), \quad 1 \leqslant j \leqslant \ell$

of the association classes of cuspidal parabolic subgroups of G.

We follow the general pattern of notation from §4. Let $L_j = \{\nu \in i\mathfrak{t}_j^*: e^\nu$ is well defined on $T_j^0\}$ and let L_j'' denote its M_j-regular set. Thus, if $\Sigma^+_{\mathfrak{t}_j}$ is a positive $\mathfrak{t}_{jC}$-root system on $\mathfrak{m}_{jC}$ and

(5.1.3a) $\tilde{\omega}_{\mathfrak{t}_j}(\nu) = \prod_{\phi \in \Sigma^+_{\mathfrak{t}_j}} \langle \phi, \nu \rangle$ for $\nu \in \mathfrak{t}^*_{jC}$,

then

(5.1.3b) $L_j'' = \{\nu \in L_j : \tilde{\omega}_{\mathfrak{t}_j}(\nu) \neq 0\}.$

Similarly, given $\zeta \in \hat{Z}$ we define

(5.1.3c) $L_{j,\zeta} = \{\nu \in L_j : e^{\nu - \rho_{\mathfrak{t}_j}}\big|_{Z \cap M_j^0} = \zeta\big|_{Z \cap M_j^0}\}$

and $L''_{j,\zeta} = L_{j,\zeta} \cap L''_j$.

Fix $\zeta \in \hat{Z}$. If $\nu \in L_j$ then $\xi = \xi_\nu$ denotes the $Z_{M_j^0}$-restriction of $e^{\nu - \rho_{\mathfrak{t}_j}}$. Proposition 4.1.6 shows that $ZZ_{M_j^0}$ is of finite index in $Z_{M_j}(M_j^0)$.

Thus we have finite subsets of $Z_{M_j}(M_j^0)^\wedge$ defined by

(5.1.4a) $S(\nu,\zeta) = Z_{M_j}(M_j^0)^\wedge_{\zeta\otimes\xi_\nu}$ if $\xi = \xi_\nu$ on $Z \cap M_j^0$, i.e., if $\nu \in L_{j,\zeta}$,

and

(5.1.4b) $S(\nu,\zeta)$ is empty if $\xi \neq \xi_\nu$ on $Z \cap M_j^0$, i.e., if $\nu \notin L_{j,\zeta}$.

Let $\nu \in L_j''$ and $\sigma \in \mathfrak{a}_j^*$. The H_j-series classes $[\pi_{\chi,\nu,\sigma}]$ that transform by $\zeta \in \hat{Z}$ are just the ones with $[\chi] \in S(\nu,\zeta)$. Thus we have finite sums

(5.1.5a) $$\pi_{j,\zeta,\nu+i\sigma} = \sum_{S(\nu,\zeta)} (\dim \chi)\pi_{\chi,\nu,\sigma},$$

(5.1.5b) $$\Theta_{j,\zeta,\nu+i\sigma} = \sum_{S(\nu,\zeta)} (\dim \chi)\Theta_{\pi_{\chi,\nu,\sigma}}.$$

If $\zeta \neq \xi_\nu$ on $Z \cap M_j^0$, i.e. if $\nu \notin L_{j,\zeta}$, then $\Theta_{j,\zeta,\nu+i\sigma} = 0$.

Here is our extension of Harish-Chandra's Plancherel formula ([25], [26]).

<u>5.1.6. Theorem</u>. *Let* $\zeta \in \hat{Z}$. *Then there are unique Borel measurable functions* $m_{j,\zeta,\nu}$ *on* $\mathfrak{a}_j^*$, $1 \leqslant j \leqslant \ell$, *with the following properties.*

1. $m_{j,\zeta,\nu}$ *are equivariant for the action of the Weyl group* W_{G,H_j}: $[wm_{j,\zeta,\nu}](\sigma) = m_{j,w^*\zeta,w^*\nu}(w^*\sigma)$.

2. $m_{j,\zeta,\nu} = 0$ *if* $\nu \notin L_{j,\zeta}$.

3. *Let* $f \in L_2(G/Z,\zeta)$ *be* C^∞ *with support compact modulo* Z. *If* $x \in G$ *define* $[r_xf](g) = f(gx)$. *Then*

(5.1.7a) $$\sum_{1\leqslant j\leqslant \ell}\ \sum_{\nu\in L''_{j,\zeta}} |\tilde{\omega}_{\mathfrak{t}_j}(\nu)| \int_{\mathfrak{a}_j^*} |\Theta_{j,\zeta,\nu+i\sigma}(r_xf)m_{j,\zeta,\nu}(\sigma)|\,d\sigma < \infty$$

and

(5.1.7b) $$f(x) = \sum_{1\leqslant j\leqslant \ell}\ \sum_{\nu\in L''_{j,\zeta}} |\tilde{\omega}_{\mathfrak{t}_j}(\nu)| \int_{\mathfrak{a}_j^*} \Theta_{j,\zeta,\nu+i\sigma}(r_xf)m_{j,\zeta,\nu}(\sigma)\,d\sigma.$$

The argument of Lemma 5.2.3 below, shows that we may drop the restriction (3.1.1) on G.

We prove Theorem 5.1.6 in §§5.2 through 5.6.

5.2. We perform some technical reductions to the proof of Theorem 5.1.6.

5.2.1. Lemma. *Let* $1 \to D \to \tilde{G} \xrightarrow{\pi} G \to 1$ *be a central extension by a discrete abelian group. Let* $\tilde{Z} = \pi^{-1}(Z)$ *and* $\tilde{\zeta} = \zeta \cdot \pi \in (\tilde{Z})^{\wedge}$. *Then* $(\tilde{G}, \tilde{Z})$ *satisfies the conditions of* §3.1, *and we may replace* (G, Z, ζ) *by* $(\tilde{G}, \tilde{Z}, \tilde{\zeta})$ *in the proof of Theorem* 5.1.6.

Proof. The formula (5.1.7) for $L_2(\tilde{G}/\tilde{Z}, \tilde{\zeta})$ is the same as the formula (5.1.7) for $L_2(G/Z, \zeta)$. *q.e.d.*

5.2.2. Lemma. *Let* W *be a closed normal abelian subgroup of* G *such that* W *centralizes* G^0 *and* $Z/Z \cap W$ *is compact. Then* (G, W) *satisfies the conditions of* §3.1, *and we may replace* (G, Z) *by* (G, W) *in the proof of Theorem* 5.1.6.

Proof. Let $Y = Z \cap W$, so Z/Y is compact. Now ZG^0/YG^0 and $(Z \cap G^0)/(Y \cap G^0)$ are compact. From the first one, since YG^0 is open in ZG^0, using (3.1.2b),

$$|G/WG^0| \leqslant |G/YG^0| = |G/ZG^0| \cdot |ZG^0/YG^0| < \infty.$$

From the second one, using (3.1.2c), $Z_{G^0}/Y \cap G^0$ is compact, so $Z_{G^0}/W \cap G^0$ is compact. Thus (G, W) satisfies (3.1.2). It follows that W/Y is compact.

Let $\nu \in L_j''$ and $\sigma \in \mathfrak{a}_j^*$. Recall (5.1.4). Let $\tau \in \hat{Y}$. If τ agrees with $\xi = \xi_\nu$ on $Y \cap M_j^0$ then

$$S(\nu, \tau) = Z_{M_j}(M_j^0)^{\wedge}_{\tau} \otimes \xi = \bigcup Z_{M_j}(M_j^0)^{\wedge}_{\zeta} \otimes \xi$$

where the union runs over $(Z\cdot Z_{M_j^0})^{\wedge}_{\tau}\otimes\xi$. Thus, in general

$$S(\nu,\tau) = \bigcup_{\zeta\in\hat{Z}_\tau} S(\nu,\zeta) \quad \text{disjoint union.}$$

Only finitely many of the $S(\nu,\zeta)$ are non-empty here because $S(\nu,\tau)$ is finite. Thus we have finite sums

$$\pi_{j,\tau,\lambda} = \sum_{\hat{Z}_\tau} \pi_{j,\zeta,\lambda} \quad\text{and}\quad \Theta_{j,\tau,\lambda} = \sum_{\hat{Z}_\tau} \Theta_{j,\zeta,\lambda}.$$

Now suppose that $f \in L_2(G/Y,\tau)$ is C^∞ with support compact modulo Y. As Z/Y is compact,

$$f = \sum_{\hat{Z}_\tau} f_\zeta \quad\text{where}\quad f_\zeta(g) = \int_{Z/Y} f(gz)\zeta(z)d(zY).$$

Here $f_\zeta \in L_2(G/Z,\zeta)$ and f_ζ is C^∞ with support compact modulo Z. Now

$$\Theta_{j,\tau,\lambda}(f) = \sum_{\zeta\in\hat{Z}_\tau} \Theta_{j,\zeta,\lambda}\Big(\sum_{\zeta'\in\hat{Z}_\tau} f_{\zeta'}\Big) = \sum_{\zeta\in\hat{Z}_\tau} \Theta_{j,\zeta,\lambda}(f_\zeta).$$

We conclude that Theorem 5.1.6 holds for $L_2(G/Y,\tau)$ if and only if, it is valid for every $L_2(G/Z,\zeta)$ with $\zeta \in \hat{Z}_\tau$. Now Theorem 5.1.6 for (G,Z) is equivalent to Theorem 5.1.6 for (G,Y). But the roles of W and Z are interchangeable because W/Y is compact. Lemma 5.2.2 is proved. *q.e.d.*

5.2.3. Lemma. *In the proof of Theorem* 5.1.6, *it suffices to consider the case where* $G = ZG^0$.

Proof. Let $\nu \in L_j''$ and $[\eta_\nu]$ the corresponding discrete class in $(M_j^0)^\wedge$. Suppose $S(\nu,\zeta)$ non-empty, i.e., that $\zeta = e^{\nu-\rho_{\mathfrak{t}_j}}$ on $Z \cap M_j^0$. Let $\sigma \in \mathfrak{a}_j^*$ and $\lambda = \nu + i\sigma$. Then

$$\begin{aligned}\pi_{j,\zeta,\lambda} &= \sum_{S(\nu,\zeta)} (\dim \chi)\pi_{\chi,\nu,\sigma} \\ &= \sum_{S(\nu,\zeta)} (\dim \chi)\operatorname{Ind}_{M_j^\dagger A_j N_j \uparrow G}(\chi \otimes \eta_\nu \otimes e^{i\sigma}) \\ &= \operatorname{Ind}_{M_j^\dagger A_j N_j \uparrow G}(\{\sum_{S(\nu,\zeta)} (\dim \chi)\chi\} \otimes \eta_\nu \otimes e^{i\sigma}).\end{aligned}$$

However, $\sum_{S(\nu,\zeta)} (\dim \chi)\chi = \operatorname{Ind}_{ZZ_{M_j^0} \uparrow Z_{M_j}(M_j^0)}(\zeta \otimes \xi_\nu)$. Thus

(5.2.4a) $\pi_{j,\zeta,\lambda} = \operatorname{Ind}_{ZM_j^0 A_j N_j \uparrow G}(\zeta \otimes \eta_\nu \otimes e^{i\sigma})$.

Adopt temporary notation $G^1 = ZG^0$, $M_j^1 = M_j \cap G^1$, $H_j^1 = H_j \cap G^1$, etc. Define $S^1(\nu,\zeta)$ as in (5.1.4), so that it consists of all $\beta \in Z_{M_j^1}(M_j^0)^\wedge$ for which we have a well defined H^1-series class $[\pi^1_{\beta,\nu,\sigma}]$ on G^1. As in (5.1.5) we have finite sums

$$\pi^1_{j,\zeta,\lambda} = \sum_{S^1(\nu,\zeta)} \pi^1_{\beta,\nu,\sigma} \quad \text{and} \quad \Theta^1_{j,\zeta,\lambda} = \sum_{S^1(\nu,\zeta)} \Theta_{\pi^1_{\beta,\nu,\sigma}} .$$

Since $S(\nu,\zeta)$ is nonempty, $S^1(\nu,\zeta)$ is nonempty, and the argument of (5.2.4a) gives

(5.2.4b) $\pi^1_{j,\zeta,\lambda} = \operatorname{Ind}_{ZM_j^0 A_j N_j \uparrow G^1}(\zeta \otimes \eta_\nu \otimes e^{i\sigma})$.

Using induction by stages now (5.2.4a, b) implies

(5.2.4c) $\pi_{j,\zeta,\lambda} = \operatorname{Ind}_{G^1 \uparrow G}(\pi^1_{j,\zeta,\lambda})$.

Note that (5.2.4c) is vacuously true when $S(\nu,\zeta)$ is empty.

Choose a system $\{x_1,\ldots,x_p\}$ of representatives of G modulo G^1. Then (5.2.4c) says that $\Theta_{j,\zeta,\lambda}$ has support in G^1 and that

$$\Theta_{j,\zeta,\lambda}(f) = \sum_{1\leqslant k\leqslant p} \Theta^1_{j,\zeta,\lambda}(f\cdot \mathrm{ad}(x_k^{-1})\big|_{G^1})$$

for $f \in L_2(G/Z,\zeta)$ which is C^∞ with support compact modulo Z. Choose $x_{jk} \in x_k G^1$ which normalize H_j, and let w_{jk} denote the element of the Weyl groups W_{G,H_j} and W_{G,A_j} given by $\mathrm{ad}(x_{jk})$. Denote $\zeta_{\pm k} = \zeta\cdot\mathrm{ad}(x_k)^{\pm 1} \in \hat{Z}$ and $\beta_{\pm k} = w_{jk}^{\pm 1}(\beta)$ for $\beta \in \mathfrak{h}^*_{jC}$. Suppose that Theorem 5.1.6 is true for G^1 with functions $m^1_{j,\zeta,\nu}$. Define

$$m_{j,\zeta,\nu}(\sigma) = \frac{1}{p^2}\sum_{1\leqslant k\leqslant p} m^1_{j,\zeta_{-k},\nu_{-k}}(\sigma_{-k}).$$

Now compute

$$f(1) = \frac{1}{p}\sum_k [f\cdot\mathrm{ad}(x_k)^{-1}](1)$$

$$= \sum_{j,k}\sum_{L''_j} |\tilde{\omega}_{\mathfrak{t}_j}(\nu)| \int_{\mathfrak{a}^*_j} \Theta^1_{j,\zeta,\nu+i\sigma}(f\cdot\mathrm{ad}(x_k^{-1})|_{G^1} \tfrac{1}{p} m^1_{j,\zeta,\nu}(\sigma)d\sigma$$

$$= \sum_{j}\sum_{L''_j} |\tilde{\omega}_{\mathfrak{t}_j}(\nu)| \int_{\mathfrak{a}^*_j} \Theta_{j,\zeta,\nu+i\sigma}(f) \tfrac{1}{p} m^1_{j,\zeta,\nu}(\sigma)d\sigma$$

$$= \sum_{j,k}\sum_{L''_j} |\tilde{\omega}_{\mathfrak{t}_j}(\nu)| \int_{\mathfrak{a}^*_j} \Theta_{j,\zeta_k,\nu_k+i\sigma_k}(f) \tfrac{1}{p^2} m^1_{j,\zeta,\nu}(\sigma)d\sigma$$

$$= \sum_{j,k}\sum_{L''_j} |\tilde{\omega}_{\mathfrak{t}_j}(\nu_{-k})| \int_{\mathfrak{a}^*_j} \Theta_{j,\zeta,\nu+i\sigma}(f) \tfrac{1}{p^2} m^1_{j,\zeta_{-k},\nu_{-k}}(\sigma_{-k})d\sigma_{-k}$$

$$= \sum_{j}\sum_{L''_j} |\tilde{\omega}_{\mathfrak{t}_j}(\nu)| \int_{\mathfrak{a}^*_j} \Theta_{j,\zeta,\nu+i\sigma}(f) m_{j,\zeta,\nu}(\sigma)d\sigma.$$

The other assertions follow.

q.e.d.

We combine Lemmas 5.2.1 through 5.2.3 for

5.2.5. Lemma. *In proving Theorem* 5.1.6, *it suffices to consider the case where*

(5.2.6a) $Z \cap G^0$ *has finite index in the center* Z_{G^0} *of* G^0,

(5.2.6b) G *and* Z *are adjusted as in Lemma* 4.3.6, *and*

(5.2.6c) $G = ZG^0$, $K = ZK^0$ *and* $K = Z(Z_{K^0})^0 \times [K^0, K^0]$.

Proof. Lemma 5.2.1 with $D = \{1\}$ or $D = Z_2$ lets us assume G adjusted as in Lemma 4.3.6. Then Lemma 5.2.1 with $D \cong (Z_{K^0})^0 \cap [K^0,K^0]$ allows us to further assume $K^0 = (Z_{K^0})^0 \times [K^0,K^0]$. Now choose a subgroup W of finite index in ZZ_{G^0} with $W(Z_{K^0})^0 \cap [K^0,K^0] = \{1\}$ and (in the notation of Lemma 4.3.6) $e^{\rho_\mathfrak{t}}(W) = 1$. Lemma 5.2.2 allows us to replace Z by W. At this point we have (5.2.6a), (5.2.6b) and $ZK^0 = Z(Z_{K^0})^0 \times [K^0,K^0]$. Finally Lemma 5.2.3 lets us suppose $G = ZG^0$. As $Z \subset K$ and $K^0 = K \cap G^0$ now $K = ZK^0$, so we have (5.2.6c). *q.e.d.*

5.3. Working in the case $G = ZG^0$, we further reduce the proof of Theorem 5.1.6 to the case where Z is the circle $S = \{s \in \mathbb{C}: |s| = 1\}$ and ζ is the character $1 = 1_S \in \hat{S}$ given by $1(s) = s$.

Assume $G = ZG^0$. Recall (3.3.1) the central extension

(5.3.1) $1 \to S \to G[\zeta] \to G/Z \to 1$ where $G[\zeta] = \{S \times G\}/\{(\zeta(z)^{-1},z): z \in Z\}$.

The projection $S \times G \to G[\zeta]$ has restriction $p: G \to G[\zeta]$. We know (Lemma 3.3.2) that $f \to f\cdot p$ is an equivariant isometry of $L_2(G[\zeta]/S,1)$ onto $L_2(G/Z,\zeta)$. Theorem 3.3.3 says that its unitary dual map

(5.3.2) $\varepsilon = \varepsilon_\zeta: G[\zeta]^\wedge_1 \to \hat{G}_\zeta$ by $\varepsilon[\psi] = [\psi\cdot p]$

is a well defined bijection that carries Plancherel measure of $G[\zeta]^\wedge_1$ to Plancherel measure of $\hat{G}_\zeta$, and that relates distribution characters by $\Theta_{\varepsilon[\psi]} = \Theta_{[\psi]}\cdot p$. We also know that ε_ζ maps $G[\zeta]^\wedge_{1\text{-disc}}$ onto $\hat{G}_{\zeta\text{-disc}}$. Given Theorem 3.3.3, it is a routine matter to verify

5.3.3. Lemma. *If* H *is a Cartan subgroup of* G *then* $H[\zeta] = \{S \times H\}/\{(\zeta(z)^{-1},z): z \in Z\}$ *is a Cartan subgroup of* $G[\zeta]$. *Further* $H \mapsto H[\zeta] = S\cdot p(H)$ *is a conjugation-equivariant bijection from the set of all Cartan subgroups of* G *to the set of all Cartan subgroups of* $G[\zeta]$. *Finally* ε_ζ *carries the* $H[\zeta]$*-series classes that transform by* $1 \in \hat{S}$ *to the* H*-series classes that transform by* $\zeta \in \hat{Z}$.

Now we must examine the action of ε_ζ on the representations and distributions of (5.1.5).

Using Lemma 5.2.5 we assume each $e^{\rho_j}(Z) = \{1\}$, so the $\mathfrak{m}_j$-regular set in $i\mathfrak{t}_j^*$ which exponentiates as in (5.1.4c) is

$$L''_{j,\zeta} = \{\nu \in i\mathfrak{t}_j^*: e^\nu \text{ defined on } T_j^0,\ e^\nu|_{Z\cap M_j^0} = \zeta|_{Z\cap M_j^0},\ \tilde{\omega}_{\mathfrak{t}_j}(\nu) \neq 0\}.$$

With that in mind set

$$L_j[\zeta]_n = \{\beta \in i\mathfrak{t}_j[\zeta]^*: e^\beta \text{ defined on } T_j[\zeta]^0 \text{ and } e^\beta(s) = s^n \text{ on } S\}$$

for every integer n. Its $\mathfrak{m}_j[\zeta]$-regular set is

$$L_j[\zeta]''_n = \{\beta \in L_j[\zeta]_n: \tilde{\omega}_{\mathfrak{t}_j}(\beta) \neq 0\}.$$

Notice that $\beta \mapsto e^\beta\cdot p$ is a bijection of $L_j[\zeta]_1$ onto $(ZT_j^0)^\wedge_\zeta = \{\zeta \otimes e^\nu: \nu \in L_{j,\zeta}\}$. Thus

(5.3.4) $e^\beta\cdot p = \zeta \otimes e^{b(\beta)}$ defines a bijection $b: L_j[\zeta]_1 \to L_{j,\zeta}$.

5.3.5. Lemma. *If* $\beta \in L_j[\zeta]_1$ *then* (*recall* (5.1.4)) $[\chi] \to [\chi\cdot p]$ *is a bijection from* $S(\beta,1)$ *to* $S(b(\beta),\zeta)$. *If* $\beta \in L_j[\zeta]''_1$ *and* $\sigma \in \mathfrak{a}_j^*$, *then* $b(\beta) \in L''_{j,\zeta}$, *each* $H_j[\zeta]$*-series class* $[\pi_{\chi,\beta,\sigma}]$ *has lift*

(5.3.6a) $[\pi_{\chi,\beta,\sigma}\cdot p] = [\pi_{\chi\cdot p,b(\beta),\sigma}]$,

and the associated distributions satisfy

(5.3.6b) $\Theta_{\pi_{\chi,\beta,\sigma}}\cdot p = \Theta_{\pi_{\chi\cdot p,b(\beta),\sigma}}$ *and* $\Theta_{j,1,\beta+i\sigma}\cdot p = \Theta_{j,\zeta,b(\beta)+i\sigma}$.

Proof. The map $\varepsilon\colon G[\zeta]_1^{\wedge} \to \hat{G}_\zeta$ restricts to a bijection $[\chi] \to [\chi\cdot p]$ of $Z_{M_j[\zeta]}(M_j[\zeta]^0)_1^{\wedge}$ onto $Z_{M_j}(M_j^0)_\zeta^{\wedge}$, and now the first assertion follows from (5.3.4). Similarly ε restricts to a bijection $M_j[\zeta]_{1\text{-disc}}^{\wedge} \to (M_j)_{\zeta\text{-disc}}^{\wedge}$ and this gives (5.3.6a). Now we have the first part of (5.3.6b), and the second follows by

$$\Theta_{j,1,\beta+i\sigma}\cdot p = \sum_{S(\beta,1)} (\dim \chi)\Theta_{\pi_{\chi,\beta,\sigma}}\cdot p$$

$$= \sum_{S(\beta,1)} (\dim(\chi\cdot p))\Theta_{\pi_{\chi\cdot p,b(\beta),\sigma}} = \Theta_{j,\zeta,b(\beta)+i\sigma}.$$

q.e.d.

We combine Lemmas 5.2.5 and 5.3.5 for

5.3.6. Lemma. *In proving Theorem* 5.1.6, *it suffices to consider the case where*

(5.3.7a) G *is connected and* Z_G *is a finite extension of a circle* S,

(5.3.7b) $(Z,\zeta) = (S,1)$ *and* (G,Z) *is adjusted as in Lemma* 4.3.6, *and*

(5.3.7c) $K = Z_K^0 \times [K,K]$ *so* $S = Z$ *is a direct factor of* K.

Proof. Lemma 5.2.5 says that we may assume (5.2.6). Then $G[\zeta]$ is connected, $Z_{G[\zeta]}$ is a finite extension of S, $(G[\zeta],S)$ is adjusted as in Lemma 4.3.6, and $K[\zeta] = Z_{K[\zeta]}^0 \times [K[\zeta],K[\zeta]]$. If Theorem 5.1.6 holds for $L_2(G[\zeta]/S,1)$ with functions $m_{j,1,\beta}$ then by Lemma 5.3.5 it holds for $L_2(G/Z,\zeta)$ with $m_{j,\zeta,b(\beta)} = m_{j,1,\beta}$.

q.e.d.

5.4. The proof of Theorem 5.1.6 was just reduced to the case (5.3.7). To further reduce it to Harish-Chandra's case we construct a certain function $E\colon G \to \mathbb{C}$.

We say that an element $g \in G$ is semisimple if $\mathrm{ad}(g)$ is a completely reducible linear transformation of $\mathfrak{g}$. The set G_{ss} of all semisimple elements of G is the union of the Cartan subgroups of G. The regular set $G' \subset G_{ss}$.

We say that an element $g \in G$ is <u>unipotent</u> if there is an Iwasawa-type decomposition $G = KAN$ with $g \in N$. That means $g = \exp(\xi)$ where $\xi \in [\mathfrak{g},\mathfrak{g}]$ such that $ad(\xi)$ is a nilpotent linear transformation. The set G_{unip} of all unipotent elements is contained in G^0.

These definitions agree with the usual ones if G is a reductive linear algebraic group. As $G/Z_G(G^0)$ is linear now, if $g \in G$ then there exist unique $g_{ss} \in G_{ss}$ and $g_u \in G_{unip}$ with $g_{ss}g_u = g = g_u g_{ss}$. We will use this notation without comment.

Now we look to the construction of functions $E: G \to C$ such that

(5.4.1a) $E(g) = E(g_{ss})$ for all $g \in G$,

(5.4.1b) $E(xgx^{-1}) = E(g)$ for all $x, g \in G$,

(5.4.1c) if $H = T \times A$ Cartan subgroup then $E|_H \in \hat{H}$ and $E(A) = \{1\}$,

(5.4.1d) if K is the fixed point set of a Cartan involution then $E|_K \in \hat{K}$.

Suppose for the moment that E satisfies (5.4.1). From the second and third conditions, E is analytic on the regular set G'. From the first and third conditions E is continuous at 1, thus by an argument of Duflo [7, Lemme 3.5] is continuous on G. In general E is not differentiable.

<u>5.4.2. Proposition</u>. *Let G be a reductive Lie group that has a Cartan involution θ. Let K be the fixed point set of θ. Then restriction $E \mapsto E|_K$ is a bijection from the set of all functions $E: G \to C$ that satisfy* (5.4.1), *to the set of all* 1-*dimensional unitary representations of* K.

Proof. Let E satisfy (5.4.1); we show that it is determined by $E|_K$. If $g \in G$ then g_{ss} is conjugate to an element (say h) of a θ-stable Cartan subgroup H of G. Split $h = ta$ along $H = T \times A$. Then $t \in K$

and $E(g) = E(g_{ss}) = E(h) = E(t)$ by (5.4.1).

Conversely let $\kappa \in \hat{K}$ be 1-dimensional; we construct E satisfying (5.4.1) with $\kappa = E|_K$. If $H = T \times A$ is a θ-stable Cartan subgroup we define $E(ta) = \kappa(t)$ for $t \in T$ and $a \in A$. If $g \in G$ then g_{ss} is conjugate to an element (say h) of a θ-stable Cartan subgroup and we define $E(g) = E(h)$. Then E is defined on G and satisfies (5.4.1). *q.e.d.*

5.4.3. Corollary. *If* G *satisfies* (5.3.7) *then there is a continuous class function* $E: G \to C$, *analytic on* G', *such that*

(5.4.4a) $E(sg) = sE(g)$ *for all* $s \in S$ *and* $g \in G$,

(5.4.4b) $E(g) = E(g_{ss})$ *for all* $g \in G$, *and*

(5.4.4c) *if* $H = T \times A$ *Cartan subgroup then* $E|_H \in \hat{H}$ *and* $E(A) = \{1\}$.

Proof. The circle S is a direct factor of K by (5.3.7c), so there exists $\kappa \in \hat{K}$ of degree 1 with $\kappa(s) = s$ for all $s \in S$. Proposition 5.4.2 provides $E: G \to C$ satisfying (5.4.1) and such that $E|_K = \kappa$. *q.e.d.*

5.5. As Lemma 5.3.6 allows, we now assume that G satisfies (5.3.7) with $(Z,\zeta) = (S,1)$. We prove an absolute continuity theorem that will be the key step in reducing the proof of Theorem 5.1.6 for $L_2(G/S,1)$ to Harish-Chandra's Plancherel Theorem for $L_2(G/S)$.

Recall (5.3.7b) that $e^{\rho_{\mathfrak{t}_j}}(S) = 1$. Now the analog of (5.1.4c) for an integer n is

(5.5.1) $L_{j,n} = \{\nu \in L_j : e^{\nu}(s) = s^n \text{ for } s \in S\}$ and $L''_{j,n} = L_{j,n} \cap L''_j$.

Choose $E: G \to C$ according to Corollary 5.4.3. Define $\varepsilon_j \in L_{j,1}$ by $e^{\varepsilon_j} = E|_{T_j^0}$. Thus $L_{j,n} = \{\beta + n\varepsilon_j : \beta \in L_{j,0}\}$.

The distributions $\Theta_{j,n,\nu+i\sigma} = \sum_{S(\nu,n)} (\dim \chi)\Theta_{\pi_{\chi,\nu,\sigma}}$, $\nu \in L''_{j,n}$ and

$\sigma \in \mathfrak{a}_j^*$, are locally integrable functions. Thus

$\Theta_{j,n,\nu+i\sigma}(f) = \int_G f(g)\Theta_{j,n,\nu+i\sigma}(g)dg$ is absolutely convergent whenever f is measurable, bounded and compactly supported.

The remainder of §5.5 is the proof of the following absolute continuity theorem.

5.5.2. Proposition. *Let* $f \in L_2(G/S,1)$ *be bounded and compactly supported. For* $1 \leqslant j \leqslant \ell$ *and* $\nu \in L''_{j,1}$ *let* $\mathfrak{b}_{j,\nu}$ *be a set of measure zero in* $\mathfrak{a}_j^*$. *Suppose*

$$\Theta_{j,1,\nu+i\sigma}(f) = 0 \quad \textit{for} \quad 1 \leqslant j \leqslant \ell,\ \nu \in L''_{j,1} \quad \textit{and} \quad \sigma \in \mathfrak{a}_j^* - \mathfrak{b}_{j,\nu}.$$

Then

$$\Theta_{j,0,\beta+i\sigma}(Ef) = 0 \quad \textit{for} \quad 1 \leqslant j \leqslant \ell,\ \beta \in L''_{j,0} \quad \textit{and} \quad \sigma \in \mathfrak{a}_j^*.$$

To prove Proposition 5.5.2 we need the explicit form of the $\Theta_{j,n,\lambda}|_{H_j \cap G'}$. Applying (5.2.4a) to our case, and noting that $S \subset T_j^0 \subset M_j^0$, we see

$$\pi_{j,n,\nu+i\sigma} = \mathrm{Ind}_{M_j^0 A_j N_j \uparrow G}(\eta_\nu \otimes e^{i\sigma}) \quad \text{for} \quad \nu \in L''_{j,n}.$$

Using induction by stages, this says

$$\pi_{j,n,\nu+i\sigma} = \mathrm{Ind}_{P_j \uparrow G}(\eta \otimes e^{i\sigma}) \quad \text{where} \quad \eta = \mathrm{Ind}_{M_j^0 \uparrow M_j}(\eta_\nu).$$

Choose $\{y_1,\ldots,y_{r_j}\}$ representatives of M_j mod M_j^0 such that T_j^0 is stable under the $\alpha_k = \mathrm{ad}(y_k)$. The distribution character Ψ_η of $\eta = \mathrm{Ind}_{M_j^0 \uparrow M_j}(\eta_\nu)$ is supported in M_j^0, and $\Psi_\eta(m) = \sum_{1 \leqslant k \leqslant r_j} \Psi_{\eta_\nu}(\alpha_k(m))$ for $m \in M_j^0$. Using (4.4.2) for Ψ_{η_ν}, now $\Psi_\eta|_{T_j \cap M_j''}$ is supported in $T_j \cap (M_j^0)''$, and

(5.5.3) $$\Psi_\eta(t) = \sum_{1\leq k\leq r_j} (-1)^{q_{M_j}(\alpha_k^* \nu)} \Delta_{M_j,T_j}(\alpha_k t)^{-1} \sum_{W_{M_j^0,T_j^0}} \det(w) e^{w\alpha_k^* \nu}$$

for $t \in T_j \cap (M_j^0)''$. Theorem 4.3.8 applies to (η,σ). In summary,

5.5.4. Lemma. *Let* $\nu \in L''_{j,n}$ *and* $\sigma \in \mathfrak{a}_j^*$. *Retain the notation above. Then* $\Theta_{j,n,\nu+i\sigma}|_{H_j \cap G'}$ *has support in* $(T_j \cap M_j^0)A_j \cap G'$. *If* $t \in T_j \cap M_j^0$ *and* $a \in A_j$ *with* $ta \in G'$, *then*

$$\Theta_{j,n,\nu+i\sigma}(ta) = \left|\frac{\Delta_{M_j,T_j}(t)}{\Delta_{G,H_j}(ta)}\right| \sum_{w(ta)\in N_G(H_j)(ta)} |N_{M_j}(T_j)(wt)|^{-1}\Psi_\eta(wt)e^{i\sigma}(wa)$$

where $\Psi_\eta(wt)$ *is given by* (5.5.3).

Our first application of the explicit form of the $\Theta_{j,n,\lambda}$ on $H_j \cap G'$ is connected with the part of Theorem 4.3.8 concerning the support of the $\Theta_{\pi_{\eta,\sigma}}$ there.

5.5.5. Proposition. *Let* f *be a bounded measurable function on* G *that has compact support contained in the closure of* G'_{H_j}. *Fix an integer* n. *For* $\nu \in L''_{j,n}$ *fix a set* $\mathfrak{b}_{j,\nu} \subset \mathfrak{a}_j^*$ *of measure zero. Suppose that*

$$\Theta_{j,n,\nu+i\sigma}(f) = 0 \quad \textit{for} \quad \nu \in L''_{j,n} \quad \textit{and} \quad \sigma \in \mathfrak{a}_j^* - \mathfrak{b}_{j,\nu}.$$

Then for $1 \leq k \leq \ell$ *we have*

$$\Theta_{k,n,\mu+i\tau}(f) = 0 \quad \textit{for} \quad \mu \in L''_{k,n} \quad \textit{and} \quad \tau \in \mathfrak{a}_k^*.$$

Proof. Partially order the Cartan subgroups of G by noncompactness as follows. $H_j \leq H_k$ means that H_k is conjugate to a subgroup of M_jA_j. Then we can replace H_k by a conjugate and assume $A_j \subset H_k$. That done, if $\sigma \in \mathfrak{a}_{jC}^*$ then $e^{i\sigma}$ is a well defined quasi-character on $H_k = (H_k \cap M_j) \times A_j$.

The role of σ in an H_j-series character $\Theta_{\pi_{\chi,\nu,\sigma}}$ is expressed by

$\Theta_{\pi_{\chi,\nu,\sigma}} = \Theta_{\pi_{\chi,\nu,0}} \cdot e^{i\sigma}$. Here we use the fact that $\Theta_{\pi_{\chi,\nu,\sigma}}$ is supported in the closure of $\bigcup_{H_j \leqslant H_k} G'_{H_k}$ and that $e^{i\sigma}$ is defined on the H_k of that union. Thus G carries invariant eigendistributions

$$\Theta_{\pi_{\chi,\nu,\sigma}} = \Theta_{\pi_{\chi,\nu,0}} \cdot e^{i\sigma} \quad \text{and} \quad \Theta_{j,n,\nu+i\sigma} = \Theta_{j,n,\nu} \cdot e^{i\sigma}$$

for $\nu \in L''_{j,n}$ and $\sigma \in \mathfrak{a}^*_{jC}$. The corresponding homomorphism $\mathfrak{Z} \to C$ is $\chi_{\nu+i\sigma}$ relative to $\mathfrak{h}_j$.

The $\Theta_{\pi_{\chi,\nu,\sigma}}(f)$, and thus the $\Theta_{j,n,\nu+i\sigma}(f)$, are visibly holomorphic in $\sigma \in \mathfrak{a}^*_{jC}$. Our hypothesis says $\Theta_{j,n,\nu+i\sigma}(f) = 0$ for $\sigma \in \mathfrak{a}^*_j - \mathfrak{b}_{j,\nu}$ where $\mathfrak{b}_{j,\nu}$ are of measure zero in $\mathfrak{a}^*_j$. We conclude $\Theta_{j,n,\nu+i\sigma}(f) = 0$ for all $\sigma \in \mathfrak{a}^*_{jC}$.

Fix k with $1 \leqslant k \leqslant \ell$. Each $\Theta_{k,n,\mu+i\tau}$ is supported in the closure of $\bigcup_{H_k \leqslant H_m} G'_{H_m}$. Since f is supported in the closure $c\ell(G'_{H_j})$, and since $c\ell(G'_{H_j}) - G'_{H_j}$ has measure zero in G, now $\Theta_{k,n,\mu+i\sigma}(f) = 0$ except possibly in the case $H_k \leqslant H_j$.

Now we may assume $H_k \leqslant H_j$. Replacing H_j by a conjugate we may suppose

$$H_j = T_j \times V_{jk} \times A_k \quad \text{with} \quad A_j = V_{jk} \times A_k \quad \text{and} \quad T_j \subset T_k.$$

Then $\mathfrak{v}^*_{jk}$ is spanned by strongly orthogonal $\mathfrak{h}_{jC}$-roots $\{\psi_\alpha\}$. Each ψ_α corresponds to a 3-dimensional simple subalgebra $\mathfrak{g}[\psi_\alpha]$ of $\mathfrak{g}$, in whose θ-stable compact real forms we have elements y_α with the following properties. First, $\theta(y_\alpha) = -y_\alpha$ and $iy_\alpha \in [\psi_\alpha]$. Second, the "partial Cayley transforms" $c_\alpha = \exp(\mathrm{ad}\, \frac{\pi}{4} y_\alpha)$ and $c_{jk} = \prod_\alpha c_\alpha$ act as the indentity on $\mathfrak{h}_j \cap \mathfrak{h}_k = \mathfrak{t}_j + \mathfrak{a}_k$ and satisfy $\mathfrak{t}_k = \mathfrak{t}_j + ic_{jk}(\mathfrak{v}_{jk})$, i.e. $\mathfrak{h}_{kC} = c_{jk}(\mathfrak{h}_{jC})$. See [45] or [54].

Let $\phi = \mu + i\tau \in i\mathfrak{h}_k^*$ be $\mathfrak{g}$-regular, where $\mu \in L''_{k,n}$ and $\tau \in \mathfrak{a}_k^*$. We claim that $\lambda = c_{jk}^*\phi$ is $\mathfrak{g}$-regular and of the form $\lambda = \nu + i\sigma$ where $\nu \in L''_{j,n}$ and $\sigma \in \mathfrak{a}_{jC}^*$. For the regularity note that c_{jk}^* maps root system to root system. For the other split $\mu = \nu + \mu_{jk}$ where $\nu \in i\mathfrak{t}_j^*$ and $\mu_{jk}(\mathfrak{t}_j) = 0$, and define $\sigma = -ic_{jk}^*(\mu_{jk}) + \tau$. Now if $z \in \mathfrak{Z}$ we compute

$$z\Theta_{k,n,\phi} = z\Theta_{j,n,c_{jk}^*\phi} = \chi_{c_{jk}^*\phi}(z)\Theta_{j,n,c_{jk}^*\phi} \text{ relative to } \mathfrak{h}_j.$$

Since $\lambda = c_{jk}^*(\phi)$ is $\mathfrak{g}$-regular, the polynomials p_s in Harish-Chandra's description [20, Theorem 4] of invariant $\mathfrak{Z}$-eigendistributions on G'_{H_j}, all are constants. Denote

$$E_{jk}(\phi) = \{\beta \in W_{\mathfrak{g}_C,\mathfrak{h}_{jC}}(\lambda)\colon e^\beta \text{ defined on } H_j^0 \text{ and } \beta|_{\mathfrak{a}_k} = i\tau\}.$$

Now Lemma 5.5.4 and Harish-Chandra's description say that $\Theta_{k,n,\phi}|_{G'_{H_j}}$ is a linear combination of the $\Theta_{j,n,\beta}|_{G'_{H_j}}$ with $\beta \in E_{j,k}(\phi)$. Our hypothesis on f was seen to force those $\Theta_{j,n,\beta}(f) = 0$. As f is supported in the closure of G'_{H_j}, we conclude that $\Theta_{k,n,\phi}(f) = 0$. *q.e.d.*

We will prove Proposition 5.5.2 by applying Proposition 5.5.5 after the following reduction.

5.5.6. Lemma. *In order to prove Proposition* 5.5.2, *it suffices to consider the case where*

(5.5.7) *if* $1 \leqslant j \leqslant \ell$ *and* $\beta \in L''_{j,0}$ *then* $\beta + \varepsilon_j \in L''_{j,1}$.

Proof. Let $\{\mathfrak{g}_\alpha\}$ be the simple ideals of $\mathfrak{g}$, $\mathfrak{k}_\alpha = \mathfrak{k} \cap \mathfrak{g}_\alpha$, and G_α and K_α the corresponding analytic subgroups of G. Each $G_\alpha \cap S = K_\alpha \cap S$ either is finite or is dense in S. Grouping the two cases we have a finite covering group $p\colon \tilde{G} \to G$, $\tilde{G} = \tilde{G}_1 \times \tilde{G}_2$, with the following properties. Let $\tilde{S}$ and $\tilde{G}_\alpha$ denote the analytic subgroups of $\tilde{G}$ for $\mathfrak{s}$ and $\mathfrak{g}_\alpha$. Then

(i) $p\colon \tilde{S} \to S$ is an isomorphism,

(ii) $\tilde{G}_1$ is generated by $\tilde{S}$ and $\{\tilde{G}_\alpha : G_\alpha \cap S$ is dense in $S\}$,

(iii) $\tilde{G}_2$ is the direct product of $\{\tilde{G}_\alpha : G_\alpha \cap S$ is finite$\}$.

Define $\tilde{E}: \tilde{G} \to C$ by $\tilde{E} = E \cdot p$. Now $\tilde{E}|_{\tilde{G}_2} = 1$, so $(\tilde{S} \times \tilde{G}_2, \tilde{E})$ satisfies (5.5.7). To prove Proposition 5.5.2 now it suffices to prove it for $(\tilde{G}_1, \tilde{E})$. Thus, in the proof of Lemma 5.5.6, we are reduced to the case where

(5.5.8) each of the $G_\alpha \cap S$ is dense in S.

We proceed by proving that (5.5.8) implies (5.5.7).

Assume (5.5.8). Then $E|_{K_\alpha}$ is a nontrivial character on K_α, for each index α. In particular G/K is a hermitian symmetric space of noncompact type. Since K has compact center (5.3.7a) now G has a compact Cartan subgroup $T \subset K$.

We prove (5.5.7) for T, i.e., for the index j such that H_j is conjugate to T. It is important to observe that the argument works for one root at a time. Let $\mathfrak{t}_\alpha = \mathfrak{t} \cap \mathfrak{k}_\alpha$, Cartan subalgebra of $\mathfrak{g}_\alpha$, so that $\mathfrak{t} = \mathfrak{s} + \sum \mathfrak{t}_\alpha$. Choose a simple $\mathfrak{t}_{\alpha C}$-root system $\{\phi_{\alpha,0}, \ldots, \phi_{\alpha,r_\alpha}\}$ for $\mathfrak{g}_{\alpha C}$ such that the $\phi_{\alpha,0}$ are noncompact roots and $\{\phi_{\alpha,1}, \ldots, \phi_{\alpha,r_\alpha}\}$ are compact. Let $\xi_{\alpha,\delta}$ be the associated basic weights, i.e. $2\langle \xi_{\alpha,\delta}, \phi_{\sigma,\tau} \rangle / \langle \phi_{\sigma,\tau}, \phi_{\sigma,\tau} \rangle = 1$ if $(\alpha,\delta) = (\sigma,\tau)$, 0 if $(\alpha,\delta) \neq (\sigma,\tau)$. Define $\varepsilon \in i\mathfrak{t}^*$ by $e^\varepsilon = E|_T$ and $\sigma \in i\mathfrak{s}^*$ by $e^\sigma(s) = s$. Since $E([K,K]) = \{1\}$, we have real numbers n_α such that $\varepsilon = \sigma + \sum n_\alpha \xi_{\alpha,0}$. Were n_α rational, the function $E|_{G_\alpha}$ would be well defined on a finite covering of the adjoint group of G_α, so $G_\alpha \cap S$ would be finite. As we are assuming (5.5.8) now each n_α is irrational. Let $\beta \in L_0'' \subset i\mathfrak{t}^*$. If ϕ is any $\mathfrak{t}_C$-root then $2\langle \beta, \phi \rangle / \langle \phi, \phi \rangle$ is a nonzero rational number. If ϕ is compact then $\langle \varepsilon, \phi \rangle = 0$ so $\langle \beta + \varepsilon, \phi \rangle = \langle \beta, \phi \rangle \neq 0$. Now let ϕ be a noncompact positive root. Then each

$$\frac{2\langle\beta+\varepsilon,\pm\phi\rangle}{\langle\phi_{\alpha,0},\phi_{\alpha,0}\rangle} = \pm\frac{\|\phi\|^2}{\|\phi_{\alpha,0}\|^2}\cdot\frac{2\langle\beta,\phi\rangle}{\langle\phi,\phi\rangle} \pm n_\alpha .$$

That is irrational because $\|\phi\|^2/\|\phi_{\alpha,0}\|^2$ and $2\langle\beta,\phi\rangle/\langle\phi,\phi\rangle$ are rational while n_α is irrational. In particular $\langle\beta + \varepsilon,\pm\phi\rangle \neq 0$. Now (5.5.7) is proved for T.

Now we prove the general case of (5.5.7). Passing to a conjugate we may suppose $T_j \subset T$. Split $\mathfrak{t} = \mathfrak{t}_j + \mathfrak{t}_j'$, direct sum that is orthogonal relative to the Killing form. As in the proof of Proposition 5.5.5 we have a partial Cayley transform $c_j \in \mathrm{Int}(\mathfrak{g}_C)$ that is the identity on $\mathfrak{t}_j$ and sends $\mathfrak{t}_j'$ to $i\mathfrak{a}_j$.

Σ denotes the $\mathfrak{t}_C$-root system on $\mathfrak{g}_C$, Σ_j denotes the $\mathfrak{h}_{jC}$-root system, and $\Sigma_{\mathfrak{t}_j} = \{\gamma \in \Sigma_j : \gamma(\mathfrak{a}_j) = 0\}$. Let $\beta \in L''_{j,0}$ and $\gamma \in \Sigma_{\mathfrak{t}_j}$. Then $2\langle\beta,\gamma\rangle/\langle\gamma,\gamma\rangle$ is a nonzero rational number and we must prove $\langle\beta + \varepsilon_j,\gamma\rangle \neq 0$. Note $\varepsilon_j = \varepsilon|_{\mathfrak{t}_j}$ where $e^{\varepsilon} = E|_T$. Since c_j is the identity on $\mathfrak{t}_j$, and since $c_j^*(\gamma)(\mathfrak{t}_j') = 0$ because $\gamma(\mathfrak{a}_j) = 0$, we have

$$\langle\beta + \varepsilon_j,\gamma\rangle = \langle\lambda + \varepsilon,\phi\rangle \quad \text{where } \lambda = c_j^*(\beta) \text{ and } \phi = c_j^*(\gamma).$$

Our argument for the case $H_j = T$ now shows $\langle\beta + \varepsilon_j,\gamma\rangle \neq 0$. *q.e.d.*

Proof of Proposition 5.5.2. According to Lemma 5.5.6 we may assume (5.5.7).

Let $f_j \in L_2(G/S,1)$ be the function that agrees with f on the closure $c\ell(G'_{H_j})$ and vanishes on $G - c\ell(G'_{H_j})$. Since the $c\ell(G'_{H_j}) - G'_{H_j}$ have measure zero in G, each

$$\Theta_{m,1,\lambda}(f) = \sum_{1\leqslant j\leqslant \ell}\int_{G'_{H_j}} f(g)\Theta_{m,1,\lambda}(g)dg = \sum_{1\leqslant j\leqslant \ell}\Theta_{m,1,\lambda}(f_j).$$

We are going to separate the summands and prove

(5.5.9) $\Theta_{k,1,\mu+i\tau}(f_j) = 0$ for $1 \leqslant j,\ k \leqslant \ell,\ \mu \in L''_{k,1}$ and $\tau \in \mathfrak{a}^*_k$

by complete induction on $\dim(T_k)$.

Fix H_k and suppose that $\Theta_{m,1,\lambda}(f_j) = 0$ whenever $\dim(T_m) < \dim(T_k)$, $\lambda \in L''_{m,1} + i\mathfrak{a}^*_m$ and $1 \leqslant j \leqslant \ell$. If j is such that $H_k \leqslant H_j$ but $H_k \neq H_j$, then $\dim(T_j) < \dim(T_k)$, so each $\Theta_{j,1,\nu+i\sigma}(f_j) = 0$. Then Proposition 5.5.5 says that the $\Theta_{k,1,\mu+i\tau}(f_j) = 0$. If j is such that $H_k \not\leqslant H_j$ then the supports of f_j and $\Theta_{k,1,\mu+i\tau}$ only meet in the set $c\ell(G'_{H_j}) - G'_{H_j}$ of measure zero in G, so $\Theta_{k,1,\mu+i\tau}(f_j) = 0$. We have seen that the $\Theta_{k,1,\mu+i\tau}(f_j) = 0$ for $j \neq k$. Finally, $\Theta_{k,1,\mu+i\tau}(f_k) = \sum_{1\leqslant j\leqslant \ell} \Theta_{k,1,\mu+i\tau}(f_j) = \Theta_{k,1,\mu+i\tau}(f) = 0$. This completes the proof of (5.5.9).

Fix k. Let $\gamma \in L''_{k,0}$ and $\tau \in \mathfrak{a}^*_k$. Since we are assuming (5.5.7) now $\mu = \gamma + \varepsilon_k$ is in $L''_{k,1}$. Thus the case $j = k$ of (5.5.9) gives

$$\Theta_{k,0,\gamma+i\tau}(Ef_k) = (E\Theta_{k,0,\gamma+i\tau})(f_k) = \Theta_{k,1,\mu+i\tau}(f_k) = 0.$$

Proposition 5.5.5 now says

$$\Theta_{j,0,\beta+i\sigma}(Ef_k) = 0 \quad \text{for } 1 \leqslant j \leqslant \ell,\ \beta \in L''_{j,0} \text{ and } \sigma \in \mathfrak{a}^*_j.$$

Finally, each

$$\Theta_{j,0,\beta+i\sigma}(Ef) = \sum_{1\leqslant k\leqslant \ell} \Theta_{j,0,\beta+i\sigma}(Ef_k) = 0.$$

This completes the proof of Proposition 5.5.2. *q.e.d.*

<u>5.6</u>. We now prove Theorem 5.1.6. As explained in Lemma 5.3.6 we work under the assumption (5.3.7) with $(Z,\zeta) = (S,1)$.

The right regular representation of G is denoted r, so

$[r_x f](g) = f(gx)$.

Choose sets $\mathfrak{b}_{j,\nu} \in \mathfrak{a}_j^*$ of Lebesgue measure zero, $1 \leqslant j \leqslant \ell$ and $\nu \in L''_{j,1}$. Consider functions $f \in C^\infty(G) \cap L_2(G/S,1)$ such that

(5.6.1a) $\Theta_{j,1,\nu+i\sigma}(r_x f) = 0$

for $x \in G$, $1 \leqslant j \leqslant \ell$, $\nu \in L''_{j,1}$ and $\sigma \in \mathfrak{a}_j^* - \mathfrak{b}_{j,\nu}$.

Then we have

(5.6.1b) $L_2(G/S,1)''$: the $L_2(G/S,1)$-closure of the space defined by (5.6.1a).

Finally we define

(5.6.1c) $L_2(G/S,1)'$: the orthocomplement of $L_2(G/S,1)''$ in $L_2(G/S,1)$.

We are going to reformulate Proposition 5.5.2 as

5.6.2. Proposition. $L_2(G/S,1)' = L_2(G/S,1)$.

Proof. Since S is compact and central in G, $L_2(G/S,1)$ is a closed bi-invariant subspace of $L_2(G/S)$. By its definition, $L_2(G/S,1)''$ is a closed right-invariant subspace of $L_2(G/S,1)$. Since the defining distributions $\Theta_{j,1,\nu+i\sigma}$ are invariant, $L_2(G/S,1)''$ also is left-invariant. Now $L_2(G/S,1)''$ is a closed bi-invariant subspace of $L_2(G)$.

Let $Q: L_2(G) \to L_2(G/S,1)''$ be orthogonal projection. If $Q \neq 0$ then I.E. Segal's argument [44, §4] gives us a function $\psi \in L_2(G/S,1)''$ and a closed nonzero subspace $V_\psi \subset L_2(G/S,1)''$ such that (i) left convolution ℓ_ψ is orthogonal projection $L_2(G) \to V_\psi$ and (ii) V_ψ is stable by every right translation r_x.

Let $\phi \in C_c^\infty(G)$, $\phi^*(x) = \overline{\phi(x^{-1})}$ as usual, and define $f = \ell_\psi(\phi^*) = \psi * \phi^*$. If $1 \leqslant j \leqslant \ell$, $[\eta_{\chi,\nu}] \in \widehat{(M_j)}_{1\text{-disc}}$ and $a \in A_j$, we define

$$F_{j,\chi,\nu}(a) = e^{-\rho_{\mathfrak{a}_j}(a)} \int_{M_j} dm \int_K dk \int_{N_j} \Psi_{\eta_{\chi,\nu}}(m) f(kmank^{-1}) dn.$$

Proposition 4.3.10 gives us the ordinary Fourier transform of $F_{j,\chi,\nu}$ *qua* function on $\mathfrak{a}_j^*$:

$$\hat{F}_{j,\chi,\nu}(\sigma) = \int_{A_j} F_{j,\chi,\nu}(a)e^{-i\sigma}(a)da = \Theta_{\pi_{\chi,\nu,-\sigma}}(f) = 0.$$

Now $F_{j,\chi,\nu} = 0$. Thus, if $\sigma \in \mathfrak{a}_{jC}^*$ then $\Theta_{\pi_{\chi,\nu,\sigma}}(f) = 0$. Now the argument of Proposition 5.5.2 applies, and so

$$\Theta_{j,0,\beta+i\sigma}(Ef) = 0 \quad \text{for} \quad 1 \leqslant j \leqslant \ell, \quad \beta \in L''_{j,0} \quad \text{and} \quad \sigma \in \mathfrak{a}_j^*.$$

Harish-Chandra's Plancherel formula for G/S ([25],[26], or see the introduction to this §5) expresses $(Ef)(1)$ as a limit of integrals of these $\Theta_{j,0,\beta+i\sigma}(Ef)$. That says

$$\langle\psi,\phi\rangle = (\psi*\phi^*)(1) = f(1) = (Ef)(1) = 0.$$

In summary, $\langle\psi,\phi\rangle = 0$ for every $\phi \in C_c^\infty(G)$.

We have just proved $\psi = 0$. Thus $V_\psi = \psi * L_2(G) = 0$; so $Q = 0$ and thus $L_2(G/S,1)'' = 0$. *q.e.d.*

Segal's general Plancherel Theorem [44] gives us a measure μ on $\hat{G}$ such that

$$f(x) = \int_{\hat{G}} \Theta_\pi(r_x f)d\mu[\pi] \quad \text{for} \quad f \in C_c^\infty(G).$$

As S is compact, $\mu_1 = \mu|_{\hat{G}_1}$ satisfies

$$f(x) = \int_{\hat{G}_1} \Theta_\pi(r_x f)d\mu_1[\pi] \quad \text{for} \quad f \in C_c^\infty(G) \cap L_2(G/S,1).$$

Proposition 5.6.2 shows that μ_1 is concentrated on the subset of $\hat{G}_1$ consisting of the irreducible constituents of the H-series classes $[\pi_{\chi,\nu,\sigma}]$, $\sigma \in \mathfrak{a}_j^* - \mathfrak{b}_{j,\nu}$. Recall (5.2.4). As $(Z,\zeta) = (S,1)$ and $S \subset M_j^0$ it says

$$\pi_{j,1,\nu+i\sigma} = \mathrm{Ind}_{M_j^0 A_j N_j \uparrow G}(\eta_\nu \otimes e^{i\sigma}).$$

It follows that the Plancherel formula for $C_c^\infty(G) \cap L_2(G/S,1)$ groups the terms $\Theta_{\pi_{\chi,\nu,\sigma}}(r_x f)$, χ variable, as a multiple of $\Theta_{j,1,\nu+i\sigma}(r_x f)$. Now

$$(5.6.3)\quad f(x) = \sum_{1 \leqslant j \leqslant \ell} \sum_{\nu \in L''_{j,1}} \int_{\mathfrak{a}_j^*} \Theta_{j,1,\nu+i\sigma}(r_x f) c_{j,\nu,\sigma} d\mu_1[\pi_{j,1,\nu+i\sigma}]$$

where $d\mu_1[\pi_{j,1,\nu+i\sigma}]$ stands for $\sum_{S(\nu,1)} (\dim \chi) d\mu_1[\pi_{\chi,\nu,\sigma}]$ and the $c_{j,\nu,\sigma}$ are constants to compensate counting each class $[\pi_{j,1,\nu+i\sigma}]$ several times. Define Borel measures $\mu_{j,1,\nu}$ on $\mathfrak{a}_j^*$ by

$$|\tilde{\omega}_{\mathfrak{t}_j}(\nu)| d\mu_{j,1,\nu}(\sigma) = c_{j,\nu,\sigma} d\mu_1[\pi_{j,1,\nu+i\sigma}].$$

Since the $\mathfrak{h}_{j,\nu}$ of definition (5.6.1) were arbitrary sets of Lebesgue measure zero in $\mathfrak{a}_j^*$, Proposition 5.6.2 shows that the measures $\mu_{j,1,\nu}$ are absolutely continuous. Let $m_{j,1,\nu}$ denote the Radon-Nikodym derivative $d\mu_{j,1,\nu}(\sigma)/d\sigma$. Then (5.6.3) becomes (5.1.7), and Theorem 5.1.6 is proved.

q.e.d.

5.7. We extract some interesting facts from the Plancherel Theorem 5.1.6.

The first consequence will be used in a later paper [57] on spaces of partially harmonic spinors.

5.7.1. Corollary. *Let* $\Omega \in \mathfrak{g}$ *be the Casimir element. If* c *is a number and* $\zeta \in \hat{Z}$ *then* $\{[\pi] \in \hat{G}_\zeta - \hat{G}_{\zeta\text{-disc}}: \chi_\pi(\Omega) = c\}$ *has Plancherel measure zero in* $\hat{G}_\zeta$.

Proof. Let $\hat{G}_{j,\zeta,c}$ denote the subset of $\hat{G}_\zeta$ consisting of the irreducible constituents of the classes $[\pi_{j,\zeta,\nu+i\sigma}]$ with $\|\nu\|^2 + \|\sigma\|^2 - \|\rho\|^2 = c$. The formula (5.1.7) and the fact $\chi_{\pi_{\chi,\nu,\sigma}}(\Omega) = \|\nu\|^2 + \|\sigma\|^2 - \|\rho\|^2$ show that

$$\{\pi \in \hat{G}_\zeta: \chi_\pi(\Omega) = c\} - \bigcup_{1 \leqslant j \leqslant \ell} \hat{G}_{j,\zeta,c}$$

has Plancherel measure zero in $\hat{G}_\zeta$. On the other hand, if $\dim \mathfrak{a}_j > 0$, then $\{\sigma \in \mathfrak{a}_j^*: \|\nu\|^2 + \|\sigma\|^2 - \|\rho\|^2 = c\}$ has Lebesgue measure zero for every $\nu \in L''_{j,\zeta}$. Now Theorem 5.1.6 says that $\hat{G}_{j,\zeta,c}$ has Plancherel measure zero.

q.e.d.

The second consequence will be used in §§7 and 8 when we consider spaces of partially harmonic (0,q)-forms.

5.7.2. Corollary. *If* G *has relative discrete series representations, and if* $\zeta \in \hat{Z}$, *then* $\{[\pi] \in \hat{G}_\zeta - \hat{G}_{\zeta\text{-disc}}: \Theta_\pi|_{K\cap G'} \neq 0\}$ *has Plancherel measure zero in* $\hat{G}_\zeta$.

This follows from the Plancherel formula and the irreducibility results stated in §4.5. In effect, Theorem 5.1.6 combines with Corollary 4.5.3 to give

5.7.3. Corollary. *Fix* $\zeta \in \hat{Z}$. *Let* $\hat{G}_{H_j,\zeta}$ *denote the set of all* H_j*-series classes* $[\pi_{\chi,\nu,\sigma}]$ *for* ζ *such that* σ *is* $(\mathfrak{g},\mathfrak{a}_j)$*-regular. Then each* $\hat{G}_{H_j,\zeta} \subset \hat{G}_\zeta$, *and the Plancherel measure of* $\hat{G}_\zeta$ *is concentrated on* $\bigcup_{1\leq j\leq \ell} \hat{G}_{H_j,\zeta}$.

If $[\pi] \in \hat{G}_{H_j,\zeta}$ with H_j/Z noncompact then $\Theta_\pi|_{K\cap G'} = 0$. Thus the set of Corollary 5.7.2 is contained in $\hat{G}_\zeta - \bigcup_{1\leq j\leq \ell} \hat{G}_{H_j,\zeta}$, and so Corollary 5.7.3 implies Corollary 5.7.2.

For completeness we give a proof of Corollary 5.7.2 that does not use the irreducibility results stated in §4.5.

5.7.4. Lemma. *Let* $[\pi]$ *be an irreducible constituent of an* H_j*-series class* $[\pi_{\chi,\nu,\sigma}]$ *where* $\nu + i\sigma \in i\mathfrak{h}_j^*$ *is* $\mathfrak{g}$*-regular. If* G *has relative discrete series representations, and if* H_j/Z *is noncompact then* $\Theta_\pi|_{K\cap G'} = 0$.

Proof. Let H/Z be a compact Cartan subgroup of G/Z. Choose an inner automorphism γ of $\mathfrak{g}_C$ such that $\gamma(\mathfrak{h}_C) = \mathfrak{h}_{jC}$. Define $\lambda = \gamma^*(\nu+i\sigma)$. Then $\lambda \in \mathfrak{h}_C^*$ is $\mathfrak{g}$-regular and $[\pi]$ has infinitesimal character χ_λ relative to $\mathfrak{h}$. Now [20, Theorem 4] there are unique constants p_w, w in the complex Weyl group $W = W_{\mathfrak{g}_C, \mathfrak{h}_C}$, such that

$\Theta_\pi|_{H^0 \cap G'} = \Delta_{G,H}^{-1} \sum p_w e^{w\lambda}$. That gives us an analytic function

(5.7.5a) $f = \Delta_{G,H} \cdot \Theta_\pi|_{H^0} = \sum_{w \in W} p_w e^{w\lambda}$ well defined on H^0.

Since λ is regular, the $w\lambda$ are distinct, so

(5.7.5b) each $p_w e^{w\lambda}$, $w \in W$, is well defined on H^0.

Let $L = \{\beta \in i\mathfrak{h}^* : e^\beta$ well defined on $H^0\}$ .and consider the Fourier expansion

$$f \sim \sum_{\beta \in L} q_\beta e^\beta .$$

If f is not identically zero, we fix an index $\beta \in L$ with $q_\beta \neq 0$. Let $\zeta^0 = e^\beta|_{Z \cap H^0}$ and consider the $L_2(H^0/Z \cap H^0, \zeta^0)$ inner product

$$0 \neq q_\beta \| e^\beta \|^2 = \langle f, e^\beta \rangle = \sum_{w \in W} p_w \langle e^{w\lambda}, e^\beta \rangle .$$

Now some $\langle e^{w\lambda}, e^\beta \rangle \neq 0$. Expressing $\langle e^{w\lambda}, e^\beta \rangle$ as a limit of integral means on $\mathfrak{h}$ we conclude $w\lambda = \beta$. This forces $w\lambda \in i\mathfrak{h}^*$. But $i\mathfrak{h}^*$ is direct sum of the real span of the $\mathfrak{h}_C$-roots with $i\mathfrak{c}^*$ where $\mathfrak{c}$ is the center of $\mathfrak{g}$. Both of those summands of $i\mathfrak{h}^*$ are W-stable. Thus every $w\lambda \in i\mathfrak{h}^*$.

Now suppose that f is not identically zero. Then we have just seen $\lambda \in i\mathfrak{h}^*$. In particular, if ϕ is an $\mathfrak{h}_C$-root then $\langle \lambda, \phi \rangle$ is real. Recall $\lambda = \gamma^*(\nu+i\sigma)$. γ^* sends the $\mathfrak{h}_{jC}$-root system to the $\mathfrak{h}_C$-root system. If

ψ is an $\mathfrak{h}_{jC}$-root now $\langle \nu+i\sigma,\psi\rangle$ is real. As $\mathfrak{a}_j^*$ is contained in the real span of the $\mathfrak{h}_{jC}$-roots now $\langle \nu+i\sigma,\mathfrak{a}_j^*\rangle = 0$, forcing $\sigma = 0$. Thus $\nu \in i\mathfrak{t}_j^*$ is a $\mathfrak{g}$-regular element of $i\mathfrak{h}_j^*$. It follows [49, Lemma 4.1] that $\mathfrak{h}_j$ is a maximally compact Cartan subalgebra of $\mathfrak{g}$. As H/Z is compact now H_j/Z is compact. That contradicts our hypothesis. Thus f is identically zero, and (5.7.5a) tells us that Θ_π vanishes on $H^0 \cap G'$.

Vanishing on $H^0 \cap G'$ forces vanishing of Θ_π on $Z_G(G^0)(H^0 \cap G') = \{Z_G(G^0)H^0\} \cap G' = H \cap G'$. As $(K \cap G') \subset G_H'$ we conclude $\Theta_\pi|_{K\cap G'} = 0$.

q.e.d.

Corollary 5.7.2 now follows from Lemma 5.7.4 and Theorem 5.1.6, using the same argument that extracted it from Corollary 5.7.3. Thus we have proved Corollary 5.7.2 in a manner independent of the irreducibility results stated in §4.5.

§6. Real Groups and Complex Flags.

G is a reductive Lie group of the class described in §3.1. Thus the adjoint representation takes G to a linear semisimple group $\bar{G} = G/Z_G(G^0)$ that has complexification

$$\bar{G}_C = \mathrm{Int}(\mathfrak{g}_C): \text{ inner automorphism group of } \mathfrak{g}_C.$$

Now G acts on all complex flag manifolds $X = \bar{G}_C/Q$. We studied the action of $\bar{G}^0$ on X in some detail [49]. Here we recall the part of [49] needed for our geometric realizations of unitary representations (cf. [50]). At the same time we extend those results from $\bar{G}^0$ to G. We also obtain some further results needed in §§7 and 8.

The notion of complex flag manifold $X = \bar{G}_C/Q$ is reviewed in §6.1. In §6.2 we recall the basic facts about the $\bar{G}$-orbit structure of X. Then we show how G acts on X in §6.3; there we digress to show how every locally compact group acts on complex flag manifolds. Open G-orbits on X are described in §6.4. In §6.5 we review the basic facts about holomorphic arc components of orbits $G(x) \subset X$. Measurable, integrable and flag type orbits are considered in §6.6. Finally, in §6.7 we give a complete analysis of a class of orbits that plays a key role in our geometric realizations of unitary representations.

Notation is changed a lot from [49] because we concentrate on the real group, its cuspidal parabolic subgroups, and the nondegenerate series representations. The main changes: the real group is denoted G, the letter P is reserved for cuspidal parabolic subgroups of G, the letter Q is used for (complex) parabolic subgroups of $\bar{G}_C$, lower case German letters denote Lie algebras, and roots are ordered so that $X = \bar{G}_C/Q$ has holomorphic tangent space spanned by positive root spaces. For convenience of the reader:

This Memoir	*References* [49] *and* [50]
$\bar{G}_C$, $\bar{\mathfrak{g}}_C$, $\bar{\mathfrak{g}}^{\phi}$, $\bar{G}$, $\bar{\mathfrak{g}}$	G, $\mathfrak{G}$, $\mathfrak{G}_{\phi}$, G_0, $\mathfrak{G}_0$
Q, $\mathfrak{q}$; Q^r, $\mathfrak{q}^r$; Q^u, $\mathfrak{q}^u$	P, $\mathcal{P}$; P^r, $\mathcal{P}^r$; P^u, $\mathcal{P}^u$
$\mathfrak{h}_C$, $\mathfrak{h}$, $\mathfrak{t} = \mathfrak{t}_H$, $\mathfrak{a} = \mathfrak{a}_H$	$\mathcal{H}$, $\mathcal{H}_0$, $\mathcal{H}_T$, $\mathcal{H}_V$
$\bar{N}_{[x]}$, $\bar{\mathfrak{n}}_{[x]}$, $\bar{N}_{[x]C}$, $\bar{\mathfrak{n}}_{[x]C}$	$N_{[x]} \cap G_0$, $\mathfrak{n}_{[x],0}$, $N_{[x]}$, $\mathfrak{n}_{[x]}$

6.1. Let $\bar{G}_C$ be a connected complex semisimple Lie group and $Q \subset \bar{G}_C$ a complex Lie subgroup. The following conditions are equivalent.

(6.1.1a) The coset space $X = \bar{G}_C/Q$ is compact.

(6.1.1b) X is a compact simply connected kaehler manifold.

(6.1.1c) X is a $\bar{G}_C$-homogeneous projective algebraic variety.

(6.1.1d) X is a closed $\bar{G}_C$-orbit in a projective representation.

(6.1.1e) Q contains a Borel subgroup of $\bar{G}_C$.

Under these conditions one says that

(6.1.2a) Q is a parabolic subgroup of $\bar{G}_C$,

(6.1.2b) $\mathfrak{q}$ is a parabolic subalgebra of $\bar{\mathfrak{g}}_C$, and

(6.1.2c) $X = \bar{G}_C/Q$ is a complex flag manifold of $\bar{G}_C$.

Given (6.1.2), Q is the analytic subgroup of $\bar{G}_C$ for $\mathfrak{q}$.

We recall the structure of parabolic subgroups and subalgebras. Choose

(6.1.3a) a Cartan subalgebra $\bar{\mathfrak{h}}_C$ of $\bar{\mathfrak{g}}_C$ and

(6.1.3b) a system Π of simple $\bar{\mathfrak{h}}_C$-roots on $\bar{\mathfrak{g}}_C$.

Then any subset $\Phi \subset \Pi$ specifies

(6.1.4a) Φ^r: all roots that are linear combinations of elements of Φ;

(6.1.4b) Φ^u: all negative roots not contained in Φ^r;

(6.1.4c) $\mathfrak{q}_{\Phi}^r = \bar{\mathfrak{h}}_C + \sum_{\Phi^r} \bar{\mathfrak{g}}^{\phi}$, $\mathfrak{q}_{\Phi}^u = \sum_{\Phi^u} \bar{\mathfrak{g}}^{\phi}$ and $\mathfrak{q}_{\Phi} = \mathfrak{q}_{\Phi}^r + \mathfrak{q}_{\Phi}^u$.

Then $\bar{G}_C$ has analytic subgroups

(6.1.5) Q_Φ^r for $\mathfrak{q}_\Phi^r$, Q_Φ^u for $\mathfrak{q}_\Phi^u$ and $Q_\Phi = Q_\Phi^r \cdot Q_\Phi^u$ for $\mathfrak{q}_\Phi$.

The situation is that

(6.1.6a) Q_Φ is a parabolic subgroup of $\bar{G}_C$;

(6.1.6b) every parabolic subgroup of $\bar{G}_C$ is conjugate to just one of the Q_Φ, $\Phi \subset \Pi$;

(6.1.6c) Q_Φ has unipotent radical Q_Φ^u and reductive part Q_Φ^r; and

(6.1.6d) Φ is a system of simple $\bar{\mathfrak{h}}_C$-roots for $\mathfrak{q}_\Phi^r$.

Let $X = \bar{G}_C/Q$ complex flag manifold. Then Q is its own normalizer in $\bar{G}_C$. Thus we have a bijective correspondence $x \leftrightarrow Q_x$ between the points of X and the $\bar{G}_C$-conjugates of Q, given by

(6.1.7) $Q_x = \{\bar{g} \in \bar{G}_C \colon \bar{g}(x) = x\}$.

We make constant use of (6.1.7) without further comment.

<u>6.2</u>. Let $\bar{G}$ be an open subgroup of a real form $\bar{G}_R$ of $\bar{G}_C$. Thus $\bar{G}^0$ is the real analytic subgroup of $\bar{G}_C$ corresponding to a real form $\bar{\mathfrak{g}} = \bar{\mathfrak{g}}_R$ of the Lie algebra $\bar{\mathfrak{g}}_C$. Denote

(6.2.1) τ: complex conjugation of $\bar{G}_C$ over $\bar{G}$, of $\bar{\mathfrak{g}}_C$ over $\bar{\mathfrak{g}}$.

Consider orbits $\bar{G}(x) \subset X = \bar{G}_C/Q$. The isotropy subgroup of $\bar{G}$ at x is $\bar{G} \cap Q_x$. The latter has Lie algebra

(6.2.2) $\bar{\mathfrak{g}} \cap \mathfrak{q}_x = \bar{\mathfrak{g}} \cap (\mathfrak{q}_x \cap \tau\mathfrak{q}_x)$, real form of $\mathfrak{q}_x \cap \tau\mathfrak{q}_x$.

The intersection of two Borel subgroups contains a Cartan subgroup. It follows that we have

(6.2.3a) a Cartan subalgebra $\bar{\mathfrak{h}} \subset \bar{\mathfrak{g}} \cap \mathfrak{q}_x$ of $\bar{\mathfrak{g}}$,

(6.2.3b) a simple system Π of $\bar{\mathfrak{h}}_C$-roots of $\bar{\mathfrak{g}}_C$, and

(6.2.3c) a subset $\Phi \subset \Pi$ such that $\mathfrak{q}_x = \mathfrak{q}_\Phi$.

Under the arrangement (6.2.3), now

(6.2.4a) $\mathfrak{q}_x \cap \tau\mathfrak{q}_x = (\mathfrak{q}_x \cap \tau\mathfrak{q}_x)^r + (\mathfrak{q}_x \cap \tau\mathfrak{q}_x)^u$ where

(6.2.4b) $(\mathfrak{q}_x \cap \tau\mathfrak{q}_x)^r = \bar{\mathfrak{h}}_C + \sum_{\Phi^r \cap \tau\Phi^r} \bar{\mathfrak{g}}^\phi$ and

(6.2.4c) $(\mathfrak{q}_x \cap \tau\mathfrak{q}_x)^u = \sum_{\Phi^r \cap \tau\Phi^u} \bar{\mathfrak{g}}^\phi + \sum_{\Phi^u \cap \tau\Phi^r} \bar{\mathfrak{g}}^\phi + \sum_{\Phi^u \cap \tau\Phi^u} \bar{\mathfrak{g}}^\phi.$

A close look at (6.2.2) and (6.2.4) shows [49, Theorem 2.1.2]

(6.2.5a) $\bar{G}(x)$ has real codimension $|\Phi^u \cap \tau\Phi^u|$ in X.

In particular,

(6.2.5b) $\bar{G}(x)$ is open in X if, and only if, $\Phi^u \cap \tau\Phi^u$ is empty.

It also follows [49, Theorem 2.6] from (6.2.2) and (6.2.4) that

(6.2.6a) there are only finitely many $\bar{G}$-orbits on X.

In effect, there are only finitely many $\bar{G}$-conjugacy classes of Cartan subalgebras $\bar{\mathfrak{h}} \subset \bar{\mathfrak{g}}$, each has only finitely many systems Π of simple roots on $\bar{\mathfrak{g}}_C$, and each $(\bar{\mathfrak{h}}, \Pi)$ corresponds to just one orbit under (6.2.3). Now, from (6.2.6a),

(6.2.6b) there are both open and closed $\bar{G}$-orbits on X.

6.3. G is a reductive Lie group of our class described in §3.1. Define $\bar{G} = G/Z_G(G^0)$. It has complexification $\bar{G}_C = \mathrm{Int}(\mathfrak{g}_C)$, and it is open in the real form $\bar{G}_R = \{\gamma \in \bar{G}_C : \gamma(\bar{\mathfrak{g}}) = \bar{\mathfrak{g}}\}$. If $X = \bar{G}_C/Q$ complex flag manifold of $\bar{G}_C$, then G acts holomorphically on X by

(6.3.1) $Q_{g(x)} = \mathrm{ad}(g)\cdot Q_x$ for $g \in G$ and $x \in X$.

This factors through the action of $\bar{G}$ on X, so

(6.3.2) if $x \in X$ then $G(x) = \bar{G}(x)$.

In particular the results of §6.2 apply to the G-orbits.

We digress to note that an arbitrary locally compact group L has holomorphic actions (6.3.1) on complex flag manifolds. Let N denote the representation radical of L^0, i.e. the intersection of the kernels of the finite dimensional irreducible complex representations of L^0. Expressing L^0 as inverse limit of Lie groups we see that L^0/N is maximal among the Lie quotient groups of L^0 that have faithful finite dimensional completely reducible representations. Let S/N denote the center of the connected reductive Lie group L^0/N. Denote

$$\tilde{Z}_L(L^0) = \{a \in L: aba^{-1} \in bS \text{ for all } b \in L^0\}.$$

In other words, $\tilde{Z}_L(L^0)/S$ is the centralizer $Z_{L/S}(L^0/S)$. Since L^0/S is a connected centerless semisimple Lie group, it is the identity component of $\bar{L} = L/\tilde{Z}_L(L^0)$, and $\bar{L}$ is a semisimple Lie group with only finitely many topological components. Were L a reductive Lie group of the class described in §3.1, then we would have $\bar{L} = L/Z_L(L^0)$ as before.

L now gives us the connected complex semisimple Lie group $\bar{L}_C = \mathrm{Int}(\bar{\mathfrak{l}}_C)$. Consider a complex flag manifold $X = \bar{L}_C/Q$. Then the action

$$Q_{a(x)} = \mathrm{ad}(\bar{a})\cdot Q_x;\ x \in X,\ a \in L \text{ and } \bar{a} = a\cdot\tilde{Z}_L(L^0),$$

is well defined precisely when the $\bar{L}_C$-conjugacy class of Q is stable under the finite group $\mathrm{ad}(\bar{L})\cdot\mathrm{Int}(\bar{\mathfrak{l}}_C)/\mathrm{Int}(\bar{\mathfrak{l}}_C)$ of outer automorphisms of $\bar{\mathfrak{l}}_C$. For example that is the case if Q is a Borel subgroup of $\bar{L}_C$. In that case, every L-orbit is a finite union of $(\bar{L})^0$-orbits, and all the results of §6.2 are valid for the L-orbit structure of X.

6.4. We return to the reductive group G of the class described in §3.1. X is a complex flag manifold $\bar{G}_C/Q$ as above. We give a rough description of the open G-orbits on X.

Let H be a Cartan subgroup of G. Choose a Cartan involution θ under which H is stable. Let $K = \{g \in G: \theta(g) = g\}$. As in (4.1.4) we decompose

$$\mathfrak{h} = \mathfrak{t} + \mathfrak{a} \quad \text{and} \quad H = T \times A \quad \text{under} \quad \theta.$$

The following conditions are equivalent [49, Lemma 4.1].

(6.4.1a) T is a Cartan subgroup of K.

(6.4.1b) $\mathfrak{t}$ contains a regular element of $\mathfrak{g}$.

(6.4.1c) Some system Π of simple $\mathfrak{h}_C$-roots has $\tau\Pi = -\Pi$.

If the conditions (6.4.1abc) hold, we say that $\mathfrak{h}$ is a maximally compact Cartan subalgebra of $\mathfrak{g}$ and that H is a maximally compact Cartan subgroup of G.

One can combine (6.2.5b) with (6.4.1) to prove [49, Theorem 4.5] that an orbit $G(x) \subset X$ is open if, and only if, there exist

(6.4.2a) a maximally compact Cartan subalgebra $\mathfrak{h} \subset \mathfrak{g}$ and

(6.4.2b) a simple $\mathfrak{h}_C$-root system Π with $\tau\Pi = -\Pi$,

such that

(6.4.2c) $\mathfrak{q}_x = \mathfrak{q}_\Phi$ for some subset $\Phi \subset \Pi$.

Further [49, Theorem 6.3] the open orbit $G(x)$ carries a G-invariant Radon measure precisely when the choices (6.4.2) can be made in such a way that $\tau\Phi^r = \Phi^r$ and $\tau\Phi^u = -\Phi^u$. In more detail, the following conditions are equivalent, and each implies that the orbit $G(x)$ be open in X.

(6.4.3a) $G(x)$ is open and has a G-invariant Radon measure.

(6.4.3b) $G(x)$ has a G-invariant indefinite-kaehler metric.

(6.4.3c) $\mathfrak{q}_x \cap \tau\mathfrak{q}_x$ is reductive, i.e. $\mathfrak{q}_x \cap \tau\mathfrak{q}_x = \mathfrak{q}_x^r \cap \tau\mathfrak{q}_x^r$.

(6.4.3d) $\mathfrak{q}_x \cap \tau\mathfrak{q}_x = \mathfrak{q}_x^r$, i.e. $\mathfrak{q}_x = \mathfrak{q}_\Phi$ with $\tau\Phi^r = \Phi^r$ and $\tau\Phi^u = -\Phi^u$.

Under the condition (6.4.3) we say that the open orbit $G(x)$ is measurable. The following conditions are equivalent [49, Theorem 6.7].

(6.4.4a) Some open G-orbit on X is measurable.

(6.4.4b) Every open G-orbit on X is measurable.

(6.4.4c) If $\mathfrak{q} = \mathfrak{q}_\Phi$ then $\tau\mathfrak{q}$ and $\mathfrak{q}^r + \mathfrak{q}^{-u}$, $\mathfrak{q}^{-u} = \sum_{\Phi^u} \bar{\mathfrak{g}}^{-\phi}$ are $\bar{G}_C$-conjugate.

These conditions are automatic [49, Corollary 6.8] if rank K = rank G, i.e. if G has relative discrete series representations. In that regard, we will need the following consequence of (6.4.2) and (6.4.3).

6.4.5. Lemma. *Let* U *be the isotropy subgroup of* G *at* $x \in X = \bar{G}_C/Q$. *Suppose that* $\mathfrak{q}$ *does not contain any nonzero ideal of* $\bar{\mathfrak{g}}_C$. *Then the following conditions are equivalent.*

(6.4.6a) U *acts on the tangent space to* G(x) *as a compact group.*

(6.4.6b) G(x) *has a* G-*invariant positive definite hermitian metric.*

(6.4.6c) $\bar{\mathfrak{g}} \cap \mathfrak{q}_x$ *is in the fixed point set of a Cartan involution of* $\bar{\mathfrak{g}}$.

Under conditions (6.4.6), G(x) *is open in* X, *and maximal compact subgroups* $\bar{K} \subset \bar{G}$ *satisfy* rank $\bar{K}$ = rank $\bar{G}$.

Suppose that rank K = rank G. Fix an open orbit G(x). Then $\bar{\mathfrak{g}}$ has a Cartan subalgebra $\bar{\mathfrak{h}} \subset \bar{\mathfrak{k}} \cap \mathfrak{q}_x$ where $\bar{H} = H/Z_G(G^0)$, $\bar{K} = K/Z_G(G^0)$ and H is a Cartan subgroup of G contained in K. Let $W_K = \{k \in \bar{K}: ad(k)\bar{\mathfrak{h}} = \bar{\mathfrak{h}}\}/\bar{H}$ and let $W_{\bar{G}_C}$ and $W_{Q_x^r}$ be the Weyl groups relative to $\bar{\mathfrak{h}}$. Then the proof of [49, Theorem 4.9] shows that the double coset space

(6.4.7) $$W_K \backslash W_{\bar{G}_C} / W_{Q_x^r}$$

enumerates the open G-orbits on X.

6.5. We look at the maximal complex analytic pieces of a general G-orbit on X.

Let V be a complex analytic space and $D \subset V$ an arbitrary subset. By holomorphic arc in D we mean a holomorphic map

$f\colon \{z \in C\colon |z| < 1\} \to V$ with image in D.

By <u>chain of holomorphic arcs</u> in D we mean a finite sequence $\{f_0, f_1, \ldots, f_m\}$ of holomorphic arcs in D such that the image of f_{i-1} meets the image of f_i $(1 \leqslant i \leqslant m)$. Now <u>holomorphic arc component</u> of D means an equivalence class of elements of D under the relation: $u \sim v$ if there is a chain $\{f_0, \ldots, f_m\}$ of holomorphic arcs in D with $u \in \text{image}(f_0)$ and $v \in \text{image}(f_m)$.

Let L be a group of holomorphic diffeomorphisms of V under which D is stable. Let S be a holomorphic arc component of D and denote its L-normalizer by

(6.5.1a) $N_L(S) = \{\alpha \in L\colon \alpha(S) = S\}$.

Then the basic observation [49, Lemma 8.2] is

(6.5.1b) $N_L(S) = \{\alpha \in L\colon \alpha(S)$ meets $S\}$,

(6.5.1c) if D is an L-orbit then S is an $N_L(S)$-orbit,

(6.5.1d) if L is a Lie transformation group on V and D is an L-orbit, then $N_L(s)$ is a Lie subgroup of L and $S \subset D \subset V$ real submanifolds.

However, in (6.5.1d) it can happen [49, §8.12] that S is not a complex submanifold of V. Here we remark that the example [49, §8.12] is correct for the quasi-split (Steinberg) form $SU(m,m)$ of $SL(2m,C)$ but false for the forms $SL(m,Q)$ and $SL(2m,R)$.

We turn to the holomorphic arc components of an orbit $G(x) \subset X = \bar{G}_C/Q$. That orbit is a finite union of G^0-orbits, so those G^0-orbits are its topological components. Holomorphic arc components are arcwise connected by construction. Thus the holomorphic arc component of $G(x) = \bar{G}(x)$ through $z \in G(x)$ coincides with the holomorphic arc component of $G^0(z) = \bar{G}^0(z)$ through z. Now there is no ambiguity in the notation

(6.5.2) $S_{[x]}$: holomorphic arc component of $G(x)$ through x.

The G-normalizer of $S_{[x]}$ is denoted

(6.5.3) $N_{[x]} = \{g \in G: gS_{[x]} = S_{[x]}\}$ with Lie algebra $\mathfrak{n}_{[x]}$.

Thus the $\bar{G}$-normalizer of $S_{[x]}$ is

$$\bar{N}_{[x]} = N_{[x]}/Z_G(G^0) \text{ with Lie algebra } \bar{\mathfrak{n}}_{[x]}.$$

The complexification $\bar{\mathfrak{n}}_{[x]C} \subset \bar{\mathfrak{g}}_C$ and we denote

(6.5.4) $\bar{N}_{[x]C}$: analytic subgroup of $\bar{G}_C$ for $\bar{\mathfrak{n}}_{[x]C}$.

The main general fact [49, Theorem 8.5 and 8.15] concerning these groups and algebras is

(6.5.5a) $\bar{N}_{[x]C}$ is a τ-stable parabolic subgroup of $\bar{G}_C$ and

(6.5.5b) $\bar{N}_{[x]}$ has finite index in the real parabolic $\bar{N}_{[x]C} \cap \bar{G}$ of $\bar{G}$.

It follows that

(6.5.5c) $N_{[x]}$ has finite index in a parabolic subgroup of G.

{Note that (6.5.5b) is weaker than [49, Theorem 8.15]. The latter has a mistake. Compare Corollary 6.7.7 with the remark just after it.}

Suppose that K is the fixed point set of a Cartan involution of G, i.e. that $\bar{K} = K/Z_G(G^0)$ is a maximal compact subgroup of $\bar{G}$. Then (6.5.5) ensures $G = KN_{[x]}$ and $\bar{G} = \bar{K}\,\bar{N}_{[x]}$. Thus K and $\bar{K}$ are transitive on

(6.5.6) $G/N_{[x]} = \bar{G}/\bar{N}_{[x]}$: space of holomorphic arc components of $G(x)$.

Fix $x \in X$ and let $\mathfrak{q}_x = \mathfrak{q}_\Phi$ as in (6.2.3). Then we have a τ-stable real linear form on $\bar{\mathfrak{h}}$ given by

(6.5.7a) $\delta_x = \sum_{\Phi^u \cap \tau\Phi^u} \phi$.

That defines a τ-stable parabolic subalgebra $\mathfrak{q}_{[x]} \subset \bar{\mathfrak{g}}_C$ by

(6.5.7b) $\mathfrak{q}_{[x]} = \bar{\mathfrak{h}}_C + \sum_{\langle\phi,\delta_x\rangle \geqslant 0} \bar{\mathfrak{g}}^\phi$.

The fact (6.5.5a) is a consequence of [49, Theorem 8.5]

(6.5.7c) $(\mathfrak{q}^u_{\mathfrak{l}x} \cap \tau\mathfrak{q}^u_{\mathfrak{l}x}) \subset \mathfrak{q}_{\mathfrak{l}[x]} \subset \{\bar{\mathfrak{n}}_{[x]C} \cap (\mathfrak{q}_{\mathfrak{l}x} + \tau\mathfrak{q}_{\mathfrak{l}x})\}$.

Now define

(6.5.8a) $\mathfrak{m}_{[x]} = \mathfrak{q}_{\mathfrak{l}[x]} + \sum_{\Gamma^0} \bar{\mathfrak{g}}^\phi$ where

(6.5.8b) $\Gamma^0 = \{\text{roots } \phi\colon \langle\phi,\delta_x\rangle < 0,\ -\phi \notin \Phi^u \cap \tau\Phi^u,\ \phi + \tau\phi \text{ not a root}\}$.

Then [49, Theorem 8.9] $\mathfrak{m}_{[x]} \subset \{\bar{\mathfrak{n}}_{[x]C} \cap (\mathfrak{q}_{\mathfrak{l}x} + \tau\mathfrak{q}_{\mathfrak{l}x})\}$ and the following conditions are equivalent.

(6.5.9a) The holomorphic arc components of $G(x)$ are complex submanifolds of X.

(6.5.9b) $\bar{\mathfrak{n}}_{[x]C} \subset \mathfrak{q}_{\mathfrak{l}x} + \tau\mathfrak{q}_{\mathfrak{l}x}$.

(6.5.9c) $\bar{\mathfrak{n}}_{[x]C}$ is equal to the space $\mathfrak{m}_{[x]}$ of (6.5.8).

(6.5.9d) $\mathfrak{m}_{[x]}$ is a subalgebra of $\bar{\mathfrak{g}}_C$.

If the conditions (6.5.9) obtain, then we say that the orbit $G(x) = \bar{G}(x) \subset X$ is <u>partially complex</u>.

<u>6.6</u>. We need stronger conditions than partial complexity for orbits. Thus we say than an orbit $G(x) \subset X$ is

(6.6.1) of <u>flag type</u> if the Zariski closure $\bar{N}_{[x]C}(x)$ of $S_{[x]} = \bar{N}_{[x]}(x)$ is a complex flag manifold;

(6.6.2) <u>measurable</u> if the holomorphic arc components of $G(x)$ carry Radon measures invariant under their normalizers; and

(6.6.3) <u>integrable</u> if $z \to \bar{\mathfrak{g}} \cap (\mathfrak{q}_{\mathfrak{l}z} + \tau\mathfrak{q}_{\mathfrak{l}z})$ is an integrable distribution on $G(x)$, i.e. if $\mathfrak{q}_{\mathfrak{l}x} + \tau\mathfrak{q}_{\mathfrak{l}x}$ is a subalgebra of $\bar{\mathfrak{g}}_C$

Let $\mathfrak{q}_{\mathfrak{l}x} = \mathfrak{q}_{\mathfrak{l}\Phi}$ as in (6.2.3) and denote

(6.6.4a) $\mathfrak{v}^-_x = \sum_{\Phi^u \cap -\tau\Phi^u} \bar{\mathfrak{g}}^\phi,\ \mathfrak{v}^+_x = \sum_{-\Phi^u \cap \tau\Phi^u} \bar{\mathfrak{g}}^\phi,\ \mathfrak{v}_x = \mathfrak{v}^-_x + \mathfrak{v}^+_x$.

Then [49, Theorem 9.2] the orbit

(6.6.4b) $G(x)$ is measurable $\Leftrightarrow \bar{\mathfrak{n}}_{[x]C} = (\mathfrak{q}_{\mathfrak{r}x} \cap \tau\mathfrak{q}_{\mathfrak{r}x}) + \mathfrak{v}_x$.

It follows that, if $G(x)$ is measurable then

(6.6.5a) the invariant Radon measure on $S_{[x]}$ comes from an $N_{[x]}$-invariant indefinite kaehler metric,

(6.6.5b) $G(x)$ is partially complex and of flag type, and

(6.6.5c) $G(x)$ is integrable if and only if $\tau\mathfrak{q}^r_{\mathfrak{r}x} = \mathfrak{q}^r_{\mathfrak{r}x}$.

On the other hand, if $\tau\mathfrak{q}^r_{\mathfrak{r}x} = \mathfrak{q}^r_{\mathfrak{r}x}$ then [49, Theorem 9.9] the following conditions are equivalent.

(6.6.6a) $G(x)$ is measurable.

(6.6.6b) $G(x)$ is integrable.

(6.6.6c) $G(x)$ is partially complex and of flag type.

Under those conditions,

(6.6.6d) $\bar{\mathfrak{n}}_{[x]C} = (\mathfrak{q}_{\mathfrak{r}x} \cap \tau\mathfrak{q}_{\mathfrak{r}x}) + \mathfrak{v}_x = \mathfrak{q}_{\mathfrak{r}[x]} = \mathfrak{q}_{\mathfrak{r}x} + \tau\mathfrak{q}_{\mathfrak{r}x}$.

Open orbits evidently are integrable, partially complex and of flag type. From either (6.4.3), or (6.6.5) and (6.6.6), an open orbit $G(x)$ is measurable just when $\tau\mathfrak{q}^r_{\mathfrak{r}x} = \mathfrak{q}^r_{\mathfrak{r}x}$.

Closed orbits also reflect some of these properties. First, we mention [49, Theorem 3.3] that

(6.6.7a) there is just one closed G-orbit on X,

(6.6.7b) it is the unique closed $\bar{G}^0$-orbit, hence is connected,

(6.6.7c) every maximal compact subgroup of $\bar{G}$ is transitive on it,

(6.6.7d) it is the unique lowest-dimensional G-orbit on X.

In [49, Theorem 9.12] we asserted that the closed orbit always is measurable, hence partially complex. That is contradicted (information received from M. Takeuchi) by applying [49, §8.12] to $G = SU(m,m)$. However, if Q is a Borel subgroup of $\bar{G}_C$, then [49, Theorem 9.12] is valid, so the closed

G-orbit is measurable, hence partially complex and of flag type. Further in that case, $\tau\mathfrak{q}^r_x = \tau\bar{\mathfrak{z}}_C = \bar{\mathfrak{z}}_C = \mathfrak{q}^r_x$, so the closed orbit is integrable by (6.6.6).

6.7. We describe a class of orbits $G(x) \subset X$ that play a key role in the geometric realization of the various nondegenerate series representations of G.

Fix a Cartan subgroup $H = T \times A$ of G and an associated cuspidal parabolic subgroup $P = MAN$ of G. We are looking for

(6.7.1a) $X = \bar{G}_C/Q$ complex flag manifold of $\bar{G}_C$ and

(6.7.1b) $Y = G(x) \subset X$ measurable integrable orbit

such that the G-normalizers of the holomorphic arc components of Y in X have the property

(6.7.1c) $N_{[x]} = \{g \in G: gS_{[x]} = S_{[x]}\}$ has Lie algebra $\mathfrak{p}$.

As the orbit $Y = G(x)$ is to be measurable, $S_{[x]}$ will be an open M^0-orbit in the sub-flag-manifold $\bar{M}_C(x)$ where $\bar{M} = M/Z_G(G^0)$. Thus AN will act trivially on $S_{[x]}$ and so the isotropy subgroup of G at x will be of the form UAN with $T \subset U \subset M$. Finally we will require that

(6.7.1d) $U/Z_G(G^0) = \{m \in M: m(x) = x\}/Z_G(G^0)$ be compact.

We first examine the algebraic consequences of (6.7.1). The Lie algebra of $Z_G(G^0)$ is the center $\mathfrak{c}$ of $\mathfrak{g}$, so we may view the Lie algebra $\bar{\mathfrak{g}}$ of $\bar{G}$ as the derived algebra of $\mathfrak{g}$; thus $\mathfrak{g} = \mathfrak{c} \oplus \bar{\mathfrak{g}}$. Let τ denote complex conjugation of $\mathfrak{g}_C$ over $\mathfrak{g}$ (as well as $\bar{\mathfrak{g}}_C$ over $\bar{\mathfrak{g}}$ and $\bar{G}_C$ over $\bar{G}$). Since $Y = G(x)$ is measurable and integrable, (6.6.5) and (6.6.6) combine with (6.7.1c) to give us

(6.7.2a) $\mathfrak{c}_C + \mathfrak{q}_x + \tau\mathfrak{q}_x = \mathfrak{p}_C$ and $\tau\mathfrak{q}^r_x = \mathfrak{q}^r_x$.

Thus the unipotent and reductive parts of $\mathfrak{p}_C$ satisfy

(6.7.2b) $\mathfrak{n}_C = \mathfrak{p}_C^u = \mathfrak{q}_x^u \cap \tau\mathfrak{q}_x^u$ and

(6.7.2c) $(\mathfrak{m}+\mathfrak{a})_C = \mathfrak{p}_C^r = (\mathfrak{c}_C + \mathfrak{q}_x^r) + (\mathfrak{q}_x^{-u} \cap \tau\mathfrak{q}_x^u) + (\mathfrak{q}_x^u \cap \tau\mathfrak{q}_x^{-u})$.

Since $\mathfrak{u} + \mathfrak{a} + \mathfrak{n} = \mathfrak{c} + (\bar{\mathfrak{q}} \cap \mathfrak{q}_x)$ we must also have

(6.7.2d) $(\mathfrak{u}+\mathfrak{a})_C = \mathfrak{c}_C + \mathfrak{q}_x^r$ and $\mathfrak{m}_C = \mathfrak{u}_C + (\mathfrak{q}_x^{-u} \cap \tau\mathfrak{q}_x^u) + (\mathfrak{q}_x^u \cap \tau\mathfrak{q}_x^{-u})$.

As $Y = G(x)$ is measurable, $S_{[x]}$ is a measurable open M^0-orbit on the flag $\bar{M}_C(x)$ where $\bar{M} = M/Z_G(G^0)$; so (6.7.2) ensures that

(6.7.3a) $\mathfrak{r} = \mathfrak{m}_C \cap (\mathfrak{c}_C + \mathfrak{q}_x)$ is a parabolic subalgebra of $\mathfrak{m}_C$

and

(6.7.3b) $\mathfrak{r}^r = \mathfrak{u}_C$ and $\mathfrak{r}^u = \mathfrak{q}_x^u \cap \tau\mathfrak{q}_x^{-u}$.

We prove that (6.7.2) and (6.7.3) characterize the situation of (6.7.1).

6.7.4. Proposition. *Let* G *be a reductive Lie group of the class described in* §3.1. *Let* $H = T \times A$ *be a Cartan subgroup and* $P = MAN$ *an associated cuspidal parabolic subgroup. Suppose that*

(i) $\mathfrak{u} \subset \mathfrak{m}$ *is the centralizer there of a subalgebra of* $\mathfrak{t}$, *such that the corresponding analytic subgroup* $U^0 \subset M^0$ *has compact image in* $M/Z_G(G^0)$;

(ii) $\mathfrak{r} \subset \mathfrak{m}_C$ *is a parabolic subalgebra with* $\mathfrak{r}^r = \mathfrak{u}_C$;

(iii) $\mathfrak{q}$ *is the* $\bar{\mathfrak{g}}_C$*-normalizer of* $\mathfrak{r}^u + \mathfrak{n}_C$ *and* Q *is the corresponding analytic subgroup of* $\bar{G}_C$;

(iv) $X = \bar{G}_C/Q$ *and* $x = 1\cdot Q \in X$.

Then Q *is a parabolic subgroup of* $\bar{G}_C$, $\mathfrak{q}^u = \mathfrak{r}^u + \mathfrak{n}_C$, *and* (X,x) *satisfies* (6.7.1). *Conversely every pair* (X,x) *satisfying* (6.7.1) *is constructed as above.*

Example. If $\mathfrak{u} = \mathfrak{t}$ then $\mathfrak{r}$ is a Borel subalgebra of $\mathfrak{m}_C$ and Q is a Borel subgroup of $\bar{G}_C$.

Proof. Denote $\bar{M} = M/Z_G(G^0)$ and let $\bar{M}_C$ be the analytic subgroup of

$\bar{G}_C$ for $\bar{\mathfrak{m}}_C = (\mathfrak{m}/\mathfrak{c})_C$. Define $\bar{R}$ to be the analytic subgroup of $\bar{M}_C$ for $\bar{\mathfrak{r}} = \mathfrak{r}/\mathfrak{c}_C$ and $S = \bar{M}_C/\bar{R}$. Then S is a complex flag manifold of $\bar{M}_C$ by (ii). Let $s = 1\cdot\bar{R} \in S$. The isotropy subalgebra of $\mathfrak{m}$ at s is $\mathfrak{c} \oplus (\bar{\mathfrak{m}} \cap \bar{\mathfrak{r}}) = \mathfrak{m} \cap \mathfrak{r}$; it has reductive part $\mathfrak{u}$ by (ii). Since $\mathfrak{t} \subset \mathfrak{u}$ by (i), (6.4.2) says that $M(s)$ is open in S and (6.4.3) says that it is measurable. Now $\mathfrak{m} \cap \mathfrak{r} = \mathfrak{u}$ and $\mathfrak{r} \cap \tau\mathfrak{r} = \mathfrak{u}_C$.

$\mathfrak{q}$ is defined (iii) as the $\bar{\mathfrak{g}}_C$-normalizer of $\mathfrak{r}^u + \mathfrak{n}_C$. The contribution to this normalizer is $\bar{\mathfrak{r}}$ from $\bar{\mathfrak{m}}_C$, $(\mathfrak{a}+\mathfrak{n})_C$ from $(\mathfrak{a}+\mathfrak{n})_C$, and 0 from $\mathfrak{n}^-_C$. Thus

$$\mathfrak{q} = (\bar{\mathfrak{u}}_C + \mathfrak{a}_C) + (\mathfrak{r}^u + \mathfrak{n}_C) \quad \text{parabolic in } \bar{\mathfrak{g}}_C.$$

Now $X = \bar{G}_C/Q$ is a complex flag manifold of $\bar{G}_C$. Further

(6.7.5a) $\mathfrak{c}_C + \mathfrak{q}^r = \mathfrak{u}_C + \mathfrak{a}_C$ and $\mathfrak{q}^u = \mathfrak{r}^u + \mathfrak{n}_C$,

in particular

(6.7.5b) $\tau\mathfrak{q}^r = \mathfrak{q}^r$ and $\mathfrak{c}_C + \mathfrak{q} + \tau\mathfrak{q} = (\mathfrak{m} + \mathfrak{a} + \mathfrak{n})_C = \mathfrak{p}_C$.

The orbit $G(x) \subset X$ is integrable now because $\mathfrak{q} + \tau\mathfrak{q} = \mathfrak{p}_C/\mathfrak{c}_C$ subalgebra of $\bar{\mathfrak{g}}_C$. As $\tau\mathfrak{q}^r = \mathfrak{q}^r$ the orbit $G(x)$ thus is measurable by (6.6.6), with $\bar{\mathfrak{n}}_{[x]C} = \mathfrak{q} + \tau\mathfrak{q} = \mathfrak{p}_C/\mathfrak{c}_C$. We conclude $\bar{\mathfrak{n}}_{[x]} = \mathfrak{p}/\mathfrak{c}$ and thus $\mathfrak{n}_{[x]} = \mathfrak{p}$. We have verified conditions (6.7.1abc) for $G(x) \subset X$.

Let U denote the isotropy subgroup of M at $x \in X$ and U_s the isotropy subgroup of M at $s \in S$. We saw that $M(s)$ is a measurable open M-orbit in S so $\mathfrak{m} \cap \mathfrak{r} = \mathfrak{u}$ and $\mathfrak{r} \cap \tau\mathfrak{r} = \mathfrak{r}^r = \mathfrak{u}_C$. Also $\mathfrak{q} \cap \tau\mathfrak{q} = (\bar{\mathfrak{u}} + \mathfrak{a} + \mathfrak{n})_C$ from (6.7.5a). Now $m(s) \to m(x)$ injects S into $G(x) \subset X$. We conclude $U = U_s$. But $U_s/Z_G(G^0)$ is compact by (i), so $U/Z_G(G^0)$ is compact, proving (6.7.1d).

In summary, the (X,x) constructed in Proposition 6.7.4 is shown to satisfy (6.7.1). The converse follows by comparison of (6.7.2) and (6.7.3) with the construction.

q.e.d.

We reformulate Proposition 6.7.4 as an enumeration of the (X,x) that satisfy (6.7.1). Choose

(6.7.6a) $\Pi_{\mathfrak{t}}$: simple $\mathfrak{t}_{\mathbb{C}}$-root system of $\mathfrak{m}_{\mathbb{C}}$

and

(6.7.6b) $\Phi_{\mathfrak{t}}$: subset of $\Pi_{\mathfrak{t}}$ consisting of compact roots.

Let Π be the unique simple $\mathfrak{h}_{\mathbb{C}}$-root system on $\mathfrak{g}_{\mathbb{C}}$ that (i) contains $\Pi_{\mathfrak{t}}$ and (ii) induces the ordering of the $\mathfrak{a}$-roots of $\mathfrak{g}$ used for construction of $P = MAN$. Define

(6.7.7c) $\Phi = \Phi_{\mathfrak{t}} \cup (\Pi - \Pi_{\mathfrak{t}}) \subset \Pi$.

The parabolic subalgebras $\mathfrak{q} \subset \bar{\mathfrak{g}}_{\mathbb{C}}$ of Proposition 6.7.4 are just the $\mathfrak{q}_{\Phi}$.

As in (3.1.3a), $M^{\dagger}$ denotes $\{m \in M: ad(m)$ is an inner automorphism on $M^0\}$.

<u>6.7.7. Corollary</u>. *If* $G(x) \subset X$ *satisfies* (6.7.1), *then* $M^{\dagger} = \{m \in M: mS_{[x]} = S_{[x]}\}$, *in particular* $U \subset M^{\dagger}$ *and* $N_{[x]} = M^{\dagger}AN$.

Remark. Lemma 8.1.3 will show that we can drop the integrability condition here.

Proof. Let $M^1 = \{m \in M: mS_{[x]} = S_{[x]}\}$. Since $M^{\dagger} = Z_M(M^0)M^0$, and $Z_M(M^0)$ acts trivially on $S_{[x]}$, we have $M^{\dagger} \subset M^1$. Thus we need only prove $M^1 \subset M^{\dagger}$.

Evidently M^1 contains the isotropy subgroup U of M at x. Further $M^1 = UM^0$ because M^0 is transitive on $S_{[x]}$. Thus we need only prove $U \subset M^{\dagger}$.

Let $u \in U$. Since $U/Z_G(G^0)$ is compact, all the Cartan subalgebras and positive Weyl chambers of $\mathfrak{u}$ are $ad(U^0)$-conjugate. Choose a positive Weyl chamber $\mathcal{C} \subset i\mathfrak{t}^*$ for $\mathfrak{u}$. Replacing u within uU^0 we may suppose $ad(u)\mathfrak{t} = \mathfrak{t}$ and $ad(u)^*\mathcal{C} = \mathcal{C}$. Since $u \in U$ and $u(x) = x$, $ad(u)$ is an inner automorphism of $\bar{\mathfrak{g}}_C$ that preserves both $\mathfrak{u}$ and $\mathfrak{q}$. From (6.7.2d) now ad(u) is trivial on (center of $\mathfrak{u}$) $\oplus\, \mathfrak{a}$. Using $\mathrm{ad}(u)^*\mathfrak{d} = \mathfrak{d}$ now ad(u) is trivial on $\mathfrak{t}$. Since T/Z is compact and $\mathrm{ad}(u) \in \mathrm{Int}(\mathfrak{m}_C)$ now $u \in T \subset M^{\dagger}$.

q.e.d.

Corollary 6.7.7 says that, if $G(x) \subset X$ satisfies (6.7.1), then $|P/N_{[x]}| = |M/M^{\dagger}|$ and P(x) has $|M/M^{\dagger}|$ topological components. Here is an example with $|M/M^{\dagger}| = 2$. Let G = SL(3,R) and let H be a maximally compact Cartan subgroup. Thus

$$\mathfrak{h} = \left\{ \begin{pmatrix} a & b & 0 \\ -b & a & 0 \\ 0 & 0 & -2a \end{pmatrix} : a, b \text{ real} \right\} \quad \text{and} \quad \mathfrak{a} = \left\{ \begin{pmatrix} a & 0 & 0 \\ 0 & a & 0 \\ 0 & 0 & -2a \end{pmatrix} : a \text{ real} \right\}.$$

Now $M = \left\{ \begin{pmatrix} A & \\ & 1/\det A \end{pmatrix} : \det A = \pm 1 \right\}$ and $M^0 \cong SL(2,R)$ given by $\det A = 1$, so $M = M^0 \cup mM^0$ where $m = \begin{pmatrix} 1 & 0 & 0 \\ 0 & -1 & 0 \\ 0 & 0 & -1 \end{pmatrix}$. One checks that ad(m) is an outer automorphism of M^0 (of course it is inner on $\mathfrak{m}_C$). Thus $M^{\dagger} = M^0$ and $|M/M^{\dagger}| = 2$. Here we refer back to the remark just after (6.5.5c).

Glancing at the argument of Corollary 6.7.7, we see that we also proved

<u>6.7.8. Corollary</u>. *If* $G(x) \subset X$ *satisfies* (6.7.1), *and if* $u \in U$, *then* ad(u) *is an inner automorphism on* U^0.

§7. Open Orbits and Discrete Series

Let G be a reductive Lie group of our class specified in §3.1. We consider complex flag manifolds $X = \bar{G}_C/Q$ and open orbits $Y = G(x) \subset X$ such that $U = \{g \in G: g(x) = x\}$ is compact modulo Z. In §7.1 we see that these pairs (X,x) exist precisely when G has relative discrete series representations, that $U = Z_G(G^0)U^0$ with $U^0 = U \cap G^0$, and that Y has $|G/G^{\dagger}|$ topological components. If $[\mu] \in \hat{U}$ we show that the associated G-homogeneous hermitian vector bundle $\mathcal{V}_\mu \to Y$ has a unique G-invariant holomorphic vector bundle structure. That allows us to construct the Hilbert spaces $H_2^{0,q}(\mathcal{V}_\mu)$ of square integrable harmonic $(0,q)$-forms on Y with values in $\mathcal{V}_\mu$, and unitary representations π_μ^q of G on $H_2^{0,q}(\mathcal{V}_\mu)$. The remainder of §7 is a fairly successful attempt to show that the $[\pi_\mu^q]$, $q \geqslant 0$ and $[\mu] \in \hat{U}$, are the relative discrete classes in $\hat{G}$.

§7.2 is the formulation and history of our main result, Theorem 7.2.3. Let $[\mu] \in \hat{U}$. Then $[\mu] = [\chi \otimes \mu^0]$ where $[\chi] \in Z_G(G^0)^\wedge$ and $[\mu^0] \in \hat{U}^0$. Let $\Theta^{disc}_{\pi_\mu^q}$ denote the character of the discrete part of π_μ^q. We prove

$$\sum_{q \geqslant 0} (-1)^q \Theta^{disc}_{\pi_\mu^q} = (-1)^{n+q(\lambda+\rho)} \Theta_{\pi_{\chi,\lambda+\rho}}$$

where λ is the highest weight of μ^0, n is the number of positive roots, and ρ is half the sum of the positive roots. We note that $H_2^{0,q}(\mathcal{V}_\mu) = 0$ for $q \neq q(\lambda + \rho)$ provided that $\lambda + \rho$ is sufficiently far from the walls of the Weyl chambers. We show that $[\pi_\mu^{q(\lambda+\rho)}] = [\pi_{\chi,\lambda+\rho}] \in \hat{G}_{disc}$ when that vanishing condition holds. Theorem 7.2.3 is proved in §§7.3 through 7.7.

We reduce the proof of Theorem 7.2.3 to the case $G = G^{\dagger}$ in §7.3, to the case $G = G^0$ in §7.4, and then further to the case where Q is a Borel subgroup of $\bar{G}_C$ in §7.5. In §7.6 we use results of Harish-Chandra and a method of W. Schmid to prove the alternating sum formula for the

$\Theta^{disc}_{\pi^q_\mu}$. The vanishing statement comes out of work of Griffiths and Schmid. When it applies, it combines with the alternating sum formula to identify $[\pi_{\chi,\lambda+\rho}] \in \hat{G}_{disc}$ as the discrete part of $[\pi^{q(\lambda+\rho)}_\mu]$. This trick is due to Narasimhan and Okamoto. Finally, we use Corollary 5.7.2 to our Plancherel Theorem to show, under condition of the vanishing statement, that $\pi^{q(\lambda+\rho)}_\mu$ has no nondiscrete part, so $[\pi^{q(\lambda+\rho)}_\mu] = [\pi_{\chi,\lambda+\rho}]$, completing our proof in §7.7.

7.1. G is a reductive Lie group of the class described in §3.1. As explained in §6.3, $\bar{G} = G/Z_G(G^0)$ has complexification $\bar{G}_C = \mathrm{Int}(\mathfrak{g}_C)$, and G acts on the complex flag manifolds of $\bar{G}_C$. To realize the relative discrete series of G we work with

(7.1.1a) $X = \bar{G}_C/Q$ complex flag manifold of $\bar{G}_C$

and

(7.1.1b) $Y = G(x) \subset X$ open G-orbit on X

such that the isotropy subgroup U of G at x satisfies

(7.1.1c) $U/Z = \{g \in G\colon g(x) = x\}/Z$ is compact.

We collect some immediate consequences of (7.1.1).

7.1.2. Lemma. *Suppose* (X,x) *given as in* (7.1.1). *Then* U/Z *contains a compact Cartan subgroup* H/Z *of* G/Z, *so* G *has relative discrete series representations. Further, the open orbit* $Y = G(x) \subset X$ *is measurable and integrable, and* (X,x) *is the case* $P = G$ *of* (6.7.1). *Finally* $U = Z_G(G^0)U^0$, $U \cap G^0 = U^0$, $UG^0 = G^\dagger$, *and* $G/G^\dagger$ *enumerates the topological components of* Y.

Remark. As consequence of the second assertion, all possibilities for (7.1.1) are enumerated in the paragraph (6.7.6) following Proposition 6.7.4.

Proof. Isotropy subgroups of G on X all contain Cartan subgroups of G by (6.2.3). Now the first assertion follows from (7.1.1c) and Theorem 3.5.8.

U acts on the tangent space as $U/Z_G(G^0)$, which is compact by (7.1.1c). Thus the orbit $Y = G(x)$ is measurable, e.g. by Lemma 6.4.5. All open orbits are integrable with $\mathfrak{q}_x + \tau\mathfrak{q}_x = \bar{\mathfrak{g}}_C$. Now we have (6.7.1) with $P = M = G = N_{[x]}$.

Let $u \in U$. Corollary 6.7.8 says that $ad(u)$ is trivial on some Cartan subalgebra of $\mathfrak{u}$, thus on a Cartan subalgebra of $\mathfrak{g}$. Now $ad(u)$ is an inner automorphism of G^0, i.e. $u \in G^\dagger$. We have just seen $U \subset G^\dagger = Z_G(G^0)G^0$. On the other hand, open orbits are simply connected, so $U \cap G^0 = U^0$. Thus $U = Z_G(G^0)U^0$ and $UG^0 = G^\dagger$. Since UG^0 is the G-normalizer of $G^0(x)$, now $G/G^\dagger$ parameterizes the components of $G(x)$.

<u>*q.e.d.*</u>

The facts about U in Lemma 7.1.2 tell us

(7.1.3a) $\hat{U} = \{[\chi \otimes \mu^0]: [\chi] \in Z_G(G^0)^\wedge \text{ and } [\mu^0] \in \hat{U}^0\}$.

In particular, since U/Z is compact,

(7.1.3b) if $[\mu] \in \hat{U}$ then its representation space V_μ has $\dim V_\mu < \infty$.

Now, for $[\mu] \in \hat{U}$ we have

(7.1.3c) $\mathcal{V}_\mu \to Y$ G-homogeneous hermitian vector bundle.

<u>7.1.4. Lemma</u>. *There is a unique complex structure on* $\mathcal{V}_\mu$ *such that* $\mathcal{V}_\mu \to Y$ *is a* G-*homogeneous holomorphic vector bundle*.

Proof. The action of G^0 on X maps $\mathfrak{g}_C$ to a Lie algebra of holomorphic vector fields. Define

$$\mathfrak{l} = \{\xi \in \mathfrak{g}_C: \xi_x = 0\},$$

isotropy subalgebra at x. The homomorphism $G^0 \to \bar{G}$ induces a homomorphism

α of $\mathfrak{g}_C$ onto $\bar{\mathfrak{g}}_C$, and $\mathfrak{l} = \alpha^{-1}(\mathfrak{q}_x)$. Note $\mathfrak{u}_C = \alpha^{-1}(\mathfrak{q}_x^r)$ reductive subalgebra of $\mathfrak{l}$. Choose a linear algebraic group with Lie algebra $\mathfrak{g}_C$ and observe that α is a homomorphism of algebraic Lie algebras. Thus $\mathfrak{u}_C$ is a maximal reductive subalgebra of $\mathfrak{l}$, and there is a nilpotent ideal $\mathfrak{l}^-$ such that $\mathfrak{l} = \mathfrak{u}_C + \mathfrak{l}^-$ semidirect sum. Observe $ad(u)\mathfrak{l}^- = \mathfrak{l}^-$ for all $u \in U$.

By extension of μ from U to $\mathfrak{l}$, we mean a (complex linear) representation λ of $\mathfrak{l}$ on V_μ such that

$$\lambda|_{\mathfrak{u}} = \mu, \text{ i.e., } \lambda(\xi) = \mu(\xi) \text{ for all } \xi \in \mathfrak{u},$$

and

$$\mu(u)\lambda(\xi)\mu(u)^{-1} = \lambda(ad(u)\xi) \text{ for all } u \in U \text{ and } \xi \in \mathfrak{l}.$$

Let λ be such an extension. Then $\lambda(\mathfrak{l}^-)$ consists of nilpotent linear transformations because λ is a representation of $\mathfrak{l}$, and that implies $\lambda(\mathfrak{l}^-) = 0$ because μ is irreducible. Thus there is just one extension of μ from U to $\mathfrak{l}$; it is given by $\lambda(\xi_1 + i\xi_2 + \eta) = \mu(\xi_1) + i\mu(\xi_2)$ where ξ_1, $\xi_2 \in \mathfrak{u}$ and $\eta \in \mathfrak{l}^-$.

Our lemma now follows from the fact [47, Theorem 3.6] that the G-homogeneous holomorphic vector bundle structures on $\mathcal{V}_\mu \to Y$ are in bijective correspondence with the extensions of μ from U to $\mathfrak{l}$. *q.e.d.*

Using (7.1.1) we fix a G-invariant hermitian metric on the complex manifold Y. The unitary structure of V_μ specifies a G-invariant hermitian metric on the fibres of $\mathcal{V}_\mu \to Y$. Denote

(7.1.5a) $A^{p,q}(\mathcal{V}_\mu) = \{C^\infty \text{ (p,q)-forms on } Y \text{ with values in } \mathcal{V}_\mu\}$.

The hermitian metrics specify the Hodge-Kodaira operators

(7.1.5b) $A^{p,q}(\mathcal{V}_\mu) \overset{\#}{\to} A^{n-p,n-q}(\mathcal{V}_\mu^*) \overset{\tilde{\#}}{\to} A^{p,q}(\mathcal{V}_\mu)$

where $n = \dim_C Y$ and $\mathcal{V}_\mu^* = \mathcal{V}_{\mu^*}$ dual bundle. If α, $\beta \in A^{p,q}(\mathcal{V}_\mu)$ then

$\alpha \wedge \#\beta \in A^{n,n}(\mathcal{V}_\mu \otimes \mathcal{V}^*_\mu)$. The pairing $V_\mu \otimes V^*_\mu \to C$ sends $\alpha \wedge \#\beta$ to an ordinary (n,n) form on Y that we denote $\alpha \barwedge \#\beta$. This gives us a pre-Hilbert space

(7.1.6a) $A_2^{p,q}(\mathcal{V}_\mu) = \{\alpha \in A^{p,q}(\mathcal{V}_\mu): \int_Y \alpha \barwedge \#\alpha < \infty\}$

whose inner product is

(7.1.6b) $\langle \alpha, \beta \rangle = \int_Y \alpha \barwedge \#\beta$.

The space of <u>square integrable</u> (p,q)-<u>forms</u> on Y with values in $\mathcal{V}_\mu$ is, by definition,

(7.1.6c) $L_2^{p,q}(\mathcal{V}_\mu)$: Hilbert space completion of $A_2^{p,q}(\mathcal{V}_\mu)$.

The operator $\bar\partial: A^{p,q}(\mathcal{V}_\mu) \to A^{p,q+1}(\mathcal{V}_\mu)$ is densely defined on $L_2^{p,q}(\mathcal{V}_\mu)$, and

(7.1.7a) $\bar\partial^* = -\tilde{\#}\,\bar\partial\,\#$ is the formal adjoint of $\bar\partial$.

That gives us a second order elliptic operator

(7.1.7b) $\square = (\bar\partial + \bar\partial^*)^2 = \bar\partial\bar\partial^* + \bar\partial^*\bar\partial$: Kodaira-Hodge-Laplacian.

The hermitian metric on Y is complete by homogeneity, so the work of Andreotti and Vesentini [1] applies. First, it says that $\square$, with domain consisting of the compactly supported forms in $A^{p,q}(\mathcal{V}_\mu)$, is essentially self adjoint on $L_2^{p,q}(\mathcal{V}_\mu)$. We also write $\square$ for the unique self adjoint extension, which coincides both with the adjoint and the closure. Its kernel

(7.1.8) $H_2^{p,q}(\mathcal{V}_\mu) = \{\omega \in L_2^{p,q}(\mathcal{V}_\mu): \square\omega = 0\}$

consists of the <u>square integrable harmonic</u> (p,q)-<u>forms</u> on Y with values in $\mathcal{V}_\mu$. $H_2^{p,q}(\mathcal{V}_\mu)$ is a closed subspace of $L_2^{p,q}(\mathcal{V}_\mu)$. It is contained in $A_2^{p,q}(\mathcal{V}_\mu)$ by ellipticity of $\square$. The Andreotti-Vesentini work also gives us an orthogonal direct sum decomposition

(7.1.9a) $L_2^{p,q}(\mathcal{V}_\mu) = \text{closure } \bar\partial L_2^{p,q-1}(\mathcal{V}_\mu) \oplus \bar\partial^* L_2^{p,q+1}(\mathcal{V}_\mu) \otimes H_2^{p,q}(\mathcal{V}_\mu)$.

$\{\bar\partial^*$ has closed range.$\}$ Here note that

(7.1.9b) $\bar\partial$ has kernel closure $\bar\partial L_2^{p,q-1}(\mathcal{V}_\mu) \oplus H_2^{p,q}(\mathcal{V}_\mu)$,

(7.1.9c) $\bar\partial^*$ has kernel $\bar\partial^* L_2^{p,q+1}(\mathcal{V}_\mu) \oplus H_2^{p,q}(\mathcal{V}_\mu)$.

Thus $H_2^{p,q}(\mathcal{V}_\mu)$ is a square integrable Dolbeault cohomology group.

The metrics and complex structures on Y and $\mathcal{V}_\mu$ are invariant under the action of G. Thus G acts on $L_2^{p,q}(\mathcal{V}_\mu)$ by a unitary representation $\tilde\pi_\mu^{p,q}$ that has a subrepresentation

(7.1.10a) $\pi_\mu^{p,q}$: unitary representation of G on $H_2^{p,q}(\mathcal{V}_\mu)$.

For convenience we also denote

(7.1.10b) π_μ^q: unitary representation of G on $H_2^{0,q}(\mathcal{V}_\mu)$.

Our program here in §7 is to represent the classes $\hat G_{disc}$ by the various π_μ^q.

<u>7.2</u>. Fix a compact Cartan subgroup H/Z of G/Z with $H \subset U$ as in Lemma 7.1.2. Choose a system Π of simple $\mathfrak{h}_C$-roots of $\mathfrak{g}_C$ such that $\mathfrak{q}_x = \mathfrak{q}_\Phi$ where $\Phi \subset \Pi$. Let Σ^+ denote the corresponding positive root system. As usual we define

$$\text{(7.2.1a)}\quad \rho = \frac{1}{2}\sum_{\phi\in\Sigma^+} \phi,\quad \Delta = \prod_{\phi\in\Sigma^+}(e^{\phi/2} - e^{-\phi/2}),\quad \tilde\omega(\cdot) = \prod_{\phi\in\Sigma^+}\langle\cdot,\phi\rangle.$$

Passing to a Z_2-extension if necessary (see Lemma 4.3.6) we may assume e^ρ and Δ well defined on H. Now let

(7.2.1b) $L = \{\lambda \in i\mathfrak{h}^*: e^\lambda \text{ defined on } H^0\}$ and $L' = \{\lambda \in L: \tilde\omega(\lambda) \neq 0\}$.

There is just one Cartan involution θ of G under which H is stable. Let $K = \{g \in G: \theta(g) = g\}$; then $H \subset U \subset K$. Now $\Sigma^+ = \Sigma_k^+ \cup \Sigma_m^+$ (disjoint) where Σ_k^+ consists of the compact positive roots $(\mathfrak{g}^\phi \subset \mathfrak{k}_C)$ and Σ_m^+ consists of the noncompact positive roots $(\mathfrak{g}^\phi \not\subset \mathfrak{k}_C)$. If $\lambda \in L'$ we have

(7.2.2) $q(\lambda) = |\{\phi \in \Sigma_k^+ : \langle \lambda,\phi\rangle < 0\}| + |\{\phi \in \Sigma_m^+ : \langle \lambda,\phi\rangle > 0\}|$.

Recall the statement of Theorem 3.5.9. Retain the notation of §7.1. Here is the main result of §7.

7.2.3. Theorem. *Let* $[\mu] \in \hat{U}$, *say* $[\mu] = [\chi \otimes \mu^0]$ *as in* (7.1.3). *Let* λ *be the highest weight of* μ^0 *for the positive* $\mathfrak{h}_C$*-root system* $\Sigma^+ \cap \Phi^r$ *of* $\mathfrak{u}_C$. *Then* $\lambda \in L$ *and* $[\mu] \in \hat{U}_\zeta$ *where* $\zeta \in \hat{Z}$ *coincides with* e^λ *on* $Z \cap G^0$ *and* $[\chi] \in Z_G(G^0)^\wedge_\zeta$. *Now assume* $\lambda + \rho \in L'$.

1. *Let* $\Theta^{disc}_{\pi^q_\mu}$ *denote the sum of the distribution characters of the irreducible* (i.e. ζ-*discrete*) *subrepresentations of* π^q_μ. *Then*

$$(7.2.4) \quad \sum_{q\geqslant 0} (-1)^q \Theta^{disc}_{\pi^q_\mu} = (-1)^{|\Sigma^+|+q(\lambda+\rho)}\Theta_{\pi_{\chi,\lambda+\rho}} .$$

2. *There is a constant* $b \geqslant 0$ *that depends only on the derived algebra* $[\mathfrak{g},\mathfrak{g}]$ *such that, if* $|\langle \lambda + \rho,\phi\rangle| > b$ *for every* $\phi \in \Sigma^+$, *then* $H_2^{0,q}(\mathcal{V}_\mu) = 0$ *whenever* $q \neq q(\lambda+\rho)$.

3. *If* $H_2^{0,q}(\mathcal{V}_\mu) = 0$ *whenever* $q \neq q_0$, *then* $[\pi^{q_0}_\mu]$ *is the* ζ-*discrete series class* $[\pi_{\chi,\lambda+\rho}]$.

Theorem 7.2.3 provides a number of explicit geometric realizations of "most" of the relative discrete series representations of G; it provides implicit geometric realizations of every class in $\hat{G}_{disc}$. In case G is a connected semisimple Lie group with finite center and $U = H$, parts 2 and 3 are due to Wilfried Schmid [43]; to some extent we follow his ideas. In case G is a connected semisimple Lie group with finite center and the orbit $Y = G(x)$ is a hermitian symmetric space, the result was proved by M.S. Narasimhan and K. Okamoto [38]. Some results for groups with possibly infinite center were proved by Harish-Chandra [19] and Juan A. Tirao [46]. Also, Schmid (unpublished) and R. Parthasarathy ([39], [40]) obtained realizations on spaces of square integrable harmonic spinors, improving the condition of part 2 for linear semisimple groups. Finally, R. Hotta [28] realized discrete series representations of connected semisimple groups of finite center on certain eigenspaces of the Casimir operator.

We carry out the proof of Theorem 7.2.3 in §§7.3 through 7.7.

7.3. We reduce Theorem 7.2.3 to the case $G = G^\dagger$.

Choose a system $\{g_1, \ldots, g_r\}$ of coset representatives of G modulo $G^\dagger$. According to Lemma 7.1.2, the topological components of $Y = G(x)$ are the $Y_i = G^\dagger(g_i x)$. Let ${}^i\pi_\mu^q$ denote the representation of $G^\dagger$ on

$$H_2^{0,q}(\mathcal{V}_\mu|_{Y_i}) = \{\omega \in H_2^{0,q}(\mathcal{V}_\mu): \ \omega \text{ supported in } Y_i\}.$$

Evidently $H_2^{0,q}(\mathcal{V}_\mu) = H_2^{0,q}(\mathcal{V}_\mu|_{Y_1}) \oplus \ldots \oplus H_2^{0,q}(\mathcal{V}_\mu|_{Y_r})$ as orthogonal direct sum, so $\pi_\mu^q|_{G^\dagger} = {}^1\pi_\mu^q \oplus \ldots \oplus {}^r\pi_\mu^q$. Also $\pi_\mu^q(g_i)$ sends $H_2^{0,q}(\mathcal{V}_\mu|_{Y_j})$ to $H_2^{0,q}(\mathcal{V}_\mu|_{Y_k})$ where $g_i Y_k = Y_j$, i.e. where $g_i^{-1} g_j \in g_k G^\dagger$. In summary

<u>7.3.1. Lemma</u>. $\pi_\mu^q = \mathrm{Ind}_{G^\dagger \uparrow G}({}^i\pi_\mu^q)$ <u>*for*</u> $1 \leqslant i \leqslant r$.

We know from Theorem 3.5.9 that $\hat{G}_{disc}$ consists of the classes $[\pi] = [\mathrm{Ind}_{G^\dagger \uparrow G}(\pi^\dagger)]$ where $[\pi^\dagger] \in (G^\dagger)_{disc}$. Further, Θ_π is supported in $G^\dagger$, where it coincides with $\Theta_{(\pi|G^\dagger)}$. Now Lemma 7.3.1 tells us that, if Theorem 7.2.3 holds for $G^\dagger$ with each of the $\mathcal{V}_\mu|_{Y_i}$, then Theorem 7.2.3 is valid for G with $\mathcal{V}_\mu$. In summary

<u>7.3.2. Lemma</u>. <u>*In the proof of Theorem*</u> 7.2.3 <u>*we may assume*</u> $G = G^\dagger$.

7.4. We reduce Theorem 7.2.3 to the case where G is connected.

Using Lemma 7.3.2 we assume $G = G^{\dagger}$. Thus $G = Z_G(G^0)G^0$ and $Y = G(x)$ is connected. Recall $[\mu] = [\chi \otimes \mu^0]$ with $[\chi] \in Z_G(G^0)^{\wedge}$ and $[\mu^0] \in \hat{U}^0$, so $V_\mu = V_\chi \otimes V_{\mu^0}$. Now $[\mu^0]$ specifies a G^0-homogeneous holomorphic vector bundle $\mathcal{V}_{\mu^0} \to Y$. Let $\pi^q_{\mu^0}$ denote the representation of G^0 on $H_2^{0,q}(\mathcal{V}_{\mu^0})$.

7.4.1 Lemma. $\pi^q_\mu = \chi \otimes \pi^q_{\mu^0}$ *for all* $q \geqslant 0$.

Proof. $Z_G(G^0)$ acts trivially on X, so it acts trivially on the bundle of ordinary $(0,q)$-forms over the orbit $Y \subset X$. Thus $Z_G(G^0)$ acts on $L_2^{0,q}(\mathcal{V}_\mu)$ as a type I primary representation $\infty\chi$. In particular $\pi^q_\mu|_{Z_G(G^0)}$ is a multiple of χ. On the other hand $\mu|_{U^0} = (\dim \chi)\mu^0$ so $\pi^q_\mu|_{G^0} = (\dim \chi)\pi^q_{\mu^0}$. We conclude $\pi^q_\mu = \chi \otimes \pi^q_{\mu^0}$, *q.e.d.*

We know from Proposition 3.5.2 that $\hat{G}_{disc}$ consists of the $[\chi' \otimes \pi^0]$ where $\chi' \in Z_G(G^0)^{\wedge}$ and $[\pi^0] \in (G^0)^{\wedge}_{disc}$ agree on Z_{G^0}. The distribution character $\Theta_{\chi' \otimes \pi^0} = (\text{trace } \chi')\Theta_{\pi^0}$. If Theorem 7.2.3 holds for G^0 with $\mathcal{V}_{\mu^0}$, now Lemma 7.4.1 ensures the result for G with $\mathcal{V}_\mu$. In summary

7.4.2. Lemma. *In the proof of Theorem* 7.2.3 *we may assume that* G *is connected*.

7.5. We reduce Theorem 7.2.3 to the case where $G = G^0$ and $U = H^0$. Choose a Borel subgroup $B \subset G$ of $\bar{G}_C$. Denote $X' = \bar{G}_C/B$ and consider the G-equivariant holomorphic projection

$$r: X' \to X \quad \text{by} \quad r(\bar{g}B) = \bar{g}Q.$$

Now choose a base point

(7.5.1a) $x' \in r^{-1}(x)$ defined by $\mathfrak{b}_{x'} = \mathfrak{q}_\emptyset$ relative to $(\mathfrak{h}_C, \Pi)$.

Since $\tau\phi = -\phi$ for every $\mathfrak{h}_{\mathbb{C}}$-root, the isotropy subalgebra of $\mathfrak{g}$ at x' is just $\mathfrak{h}$. Now

(7.5.1b) $Y' = G(x')$ is open in X' and $H = \{g \in G: g(x') = x'\}$

and

(7.5.1c) $r: Y' \to Y$ is G-equivariant and holomorphic.

Following Lemma 7.4.2, we assume G connected, so U and H are connected. Now λ is the highest weight of μ and $e^{\lambda} \in \hat{H}$ specifies

(7.5.2a) $\mathcal{L}_{\lambda} \to Y'$ G-homogeneous holomorphic line bundle.

Finally denote

(7.5.2b) π_{λ}^{q}: representation of G on $H_2^{0,q}(\mathcal{L}_{\lambda})$.

<u>7.5.3. Lemma</u>. $[\pi_{\lambda}^{q}] = [\pi_{\mu}^{q}]$.

Proof. This is a Leray spectral sequence argument.

Let $\underline{\underline{O}}(\mathcal{L}_{\lambda})$ denote the sheaf of germs of holomorphic sections of $\mathcal{L}_{\lambda} \to Y'$. Each integer $s \geqslant 0$ gives a sheaf $\underline{\underline{R}}^{s}(\mathcal{L}_{\lambda}) \to Y$, associated to the presheaf that assigns the sheaf cohomology group $H^{s}(Y' \cap r^{-1}D, \underline{\underline{O}}(\mathcal{L}_{\lambda}))$ to an open set $D \subset Y$. Since $r: Y' \to Y$ is a holomorphic fibre bundle, $\underline{\underline{R}}^{s}(\mathcal{L}_{\lambda})$ is the sheaf of germs of holomorphic section to a holomorphic vector bundle over Y whose fibre at $y \in Y$ is $H^{s}(Y' \cap r^{-1}(y), \underline{\underline{O}}(\mathcal{L}_{\lambda}))$. Recall our extension of the Borel-Weil Theorem given by Proposition 1.1.12 with $q_0 = 0$, and apply it to $Y' \cap r^{-1}(x) = U(x') \cong U/H$. That says $H^{0}(Y' \cap r^{-1}(x), \underline{\underline{O}}(\mathcal{L}_{\lambda})) = V_{\mu}$ as U-module and $H^{s}(Y' \cap r^{-1}(x), \underline{\underline{O}}(\mathcal{L}_{\lambda})) = 0$ for $s > 0$. Now $\underline{\underline{R}}^{0}(\mathcal{L}_{\lambda}) = \underline{\underline{O}}(\mathcal{V}_{\mu})$ and $\underline{\underline{R}}^{s}(\mathcal{L}_{\lambda}) = 0$ for $s > 0$.

Our analysis of the direct image sheaves $\underline{\underline{R}}^{s}(\mathcal{L}_{\lambda})$ shows that the Leray spectral sequence collapses for $r: Y' \to Y$, so each $H^{q}(Y', \underline{\underline{O}}(\mathcal{L}_{\lambda})) = H^{q}(Y, \underline{\underline{O}}(\mathcal{V}_{\mu}))$ as G-modules. More to the point, we carry the spectral sequence over from sheaf cohomology to Dolbeault cohomology and use the Andreotti-Vesentini Theorem (7.1.9) to restrict considerations to square integrable forms. Then the resultant square integrable Leray spectral sequence collapses and we conclude that each $H_2^{0,q}(Y', \mathcal{L}_\lambda) = H_2^{0,q}(Y, \mathcal{V}_\mu)$ as G-modules. *q.e.d.*

As immediate consequence of Lemmas 7.4.2 and 7.5.3 we have

7.5.4. Lemma. *In the proof of Theorem* 7.2.3 *we may assume that* G *is connected*, Q *is a Borel subgroup of* $\bar{G}_C$ *and* $U = H$.

7.6. We prove the alternating sum formula (7.2.4). In view of Lemma 7.5.4 we assume G connected, Q Borel in $\bar{G}_C$ and $U = H$.

K/Z is the maximal compact subgroup of G/Z that contains the compact Cartan subgroup H/Z. If $[\pi] \in \hat{G}_\zeta$ we know (3.2.2) that

(7.6.1a) $\pi|_K = \sum_{\hat{K}_\zeta} m_\kappa \cdot \kappa, \quad 0 \leqslant m_\kappa \leqslant n_G(\dim \kappa)$,

for some integer $n_G > 0$. Thus Harish-Chandra's argument [15, §5] says that each

$$(\pi|_K)(f) = \int_K f(k)\pi(k)dk, \quad f \in C_c^\infty(K),$$

is of trace class and that

(7.6.1b) $T_\pi : C_c^\infty(K) \to C$ by $f \mapsto \mathrm{trace}(\pi|_K)(f)$

is a distribution on K. As usual let G' denote the regular set in G. Harish-Chandra's argument [20, §12] now shows that $T_\pi|_{K \cap G'}$ is a real analytic function on $K \cap G'$ and that

(7.6.1c) $T_\pi|_{K\cap G'} = \Theta_\pi|_{K \cap G'}$.

θ is the Cartan involution of G with fixed point set K. Fix a non-degenerate invariant bilinear form $\langle\ ,\ \rangle$ on $\mathfrak{g}_C$ that restricts to the Killing form on the derived algebra and is negative definite on $\mathfrak{k} = (\mathfrak{k} \cap [\mathfrak{g},\mathfrak{g}]) \oplus \mathfrak{c}$. That gives us a positive definite ad(K)-invariant hermitian inner product

(7.6.2) $(u,v) = -\langle u,\theta\tau v\rangle$

on $\mathfrak{g}_C$ where τ is complex conjugation over $\mathfrak{g}$.

Consider the nilpotent algebra

(7.6.3a) $\mathfrak{n} = \sum_{\phi\in\Sigma^+} \bar{\mathfrak{g}}^{-\phi} = \mathfrak{q}_x^u \subset \bar{\mathfrak{g}}_C$.

Denote

(7.6.3b) ad^*: representation of $\mathfrak{q}_x$ on $\Lambda\mathfrak{n}^* = \sum_{j\geqslant 0} \Lambda^j\mathfrak{n}^*$.

The inner product (7.6.2) gives $\mathfrak{n}$, thus $\mathfrak{n}^*$, thus also $\Lambda\mathfrak{n}^*$, a Hilbert space structure; and $ad^*(\mathfrak{h})$ acts by skew-hermitian transformations.

Fix $[\pi] \in \hat{G}_\zeta$ and let H_π^0 denote the space of K-finite vectors in H_π. It is dense and consists of analytic vectors (7.6.1a). Now

$$\mathfrak{h} \text{ acts on } H_\pi^0 \otimes \Lambda\mathfrak{n}^* \text{ by } \pi \otimes ad^*,$$

and the action is by skew-hermitian transformations. Let $\{y_1,\ldots,y_n\}$ basis of $\mathfrak{n}$, $\{\omega^j\}$ the dual basis of $\mathfrak{n}^*$, and $e(\omega^j)$: $\Lambda\mathfrak{n}^* \to \Lambda\mathfrak{n}^*$ exterior product. Then

$$\delta = \sum\{\pi(y_j) \otimes e(\omega^j) + \frac{1}{2}\, 1 \otimes e(\omega^j)ad^*(y_j)\}$$

is the coboundary operator $H_\pi^0 \otimes \Lambda\mathfrak{n}^* \to H_\pi^0 \otimes \Lambda\mathfrak{n}^*$ of the Lie algebra cohomology for the action of $\mathfrak{h}$. It has formal adjoint

$$\delta^* = \sum\{-\pi(\tau y_j) \otimes i(\omega^j) + \frac{1}{2}\, 1 \otimes ad^*(y_j)^* i(\omega^j)\}$$

where $i(\omega^j)$ denotes interior product. Now $\delta + \delta^*$ is a densely defined symmetric operator on $H_\pi \otimes \Lambda\mathfrak{n}^*$.

Choose a basis $\{z_i\}$ of $\mathfrak{k}_C$ that is orthonormal relative to $\langle\ ,\ \rangle$. Then

$$\Omega_K = \sum z_i z_i \in \mathfrak{K}$$

is independent of choice of basis $\{z_i\}$. In particular Ω_K is a linear combination, positive coefficients, of the Casimir operators of the simple ideals of $\mathfrak{k}$ plus the Laplacian of the center of $\mathfrak{k}$. Thus (7.6.1a) $\pi(\Omega_K)$ is symmetric non-negative on H_π^0 and has a unique self adjoint extension $\pi(\Omega_K)$ to H_π. Further, H_π is the discrete direct sum of the (all non-negative) eigenspaces of $\pi(\Omega_K)$. As

$$\{[\kappa] \in \hat{K}_\zeta : \kappa(\text{Casimir element of } \mathfrak{K}) \leqslant c\}$$

is finite for every real c, (7.6.1a) also says that the sum of the eigenspaces of $\pi(\Omega_K)$ for eigenvalue $\leqslant c$ is finite. Thus

$$(1 + \pi(\Omega_K))^{-1} \text{ is a self adjoint compact operator on } H_\pi .$$

With the preparation above, Wilfried Schmid's arguments [43, §3] are valid in our situation. We state the result.

7.6.4. Lemma. *There is a unique self adjoint extension of* $\delta + \delta^*$ *on* $H_\pi \otimes \Lambda \mathfrak{n}^*$ (*it is the closure from the domain* $H_\pi^0 \otimes \Lambda\mathfrak{n}^*$). *Each*

$$\mathcal{H}^q(\pi): \textit{kernel of } \delta + \delta^* \textit{ on } H_\pi \otimes \Lambda^q \mathfrak{n}^*$$

is a finite dimensional H-*module. Define*

$$f_\pi = \sum (-1)^q (\text{character of } H \text{ on } \mathcal{H}^q(\pi)).$$

Let Δ *and* ρ *be as in* (7.2.1a) *and* $n = \dim_{\mathbb{C}} \mathfrak{n} = |\Sigma^+|$. *Then*

$$f_\pi|_{H \cap G'} = (-1)^n \Delta e^\rho \cdot T_\pi|_{H \cap G'} .$$

Let $d\pi$ denote Plancherel measure on $\hat{G}_\zeta$, so $L_2(G/Z,\zeta) = \int_{\hat{G}_\zeta} H_\pi \otimes H_\pi^* d\pi$.

We have the unitary G-module structure

$$L_2^{0,q}(\mathcal{L}_\lambda) = \int_{\hat{G}_\zeta} H_\pi \otimes \{H_\pi^* \otimes \Lambda^q \mathfrak{n}^* \otimes L_\lambda\}^H d\pi$$

where L_λ is the representation space of e^λ, H acts on $H_\pi^* \otimes \Lambda^q\mathfrak{n}^* \otimes L_\lambda$ by $\pi^* \otimes \mathrm{ad}^* \otimes e^\lambda$, and $\{\ \}^H$ denotes the fixed points of H there. Now $\bar\partial : A^{0,q}(\mathcal{L}_\lambda) \to A^{0,q+1}(\mathcal{L}_\lambda)$ and its formal adjoint $\bar\partial^*$ act by

$$\bar\partial(f \cdot \omega^J \cdot \ell) = \sum_{1 \leqslant k \leqslant n} y_k(f) \cdot e(\omega^k)\omega^J \cdot \ell + \frac{1}{2} \sum_{1 \leqslant k \leqslant n} f \cdot e(\omega^k)\mathrm{ad}^*(y_k)\omega^J \cdot \ell$$

and

$$\bar\partial^*(f\cdot\omega^I\cdot\ell) = -\sum_{1\leq k\leq n}\tau(y_k)f\cdot i(\omega^k)\omega^I\cdot\ell + \frac{1}{2}\sum_{1\leq k\leq n} f\cdot ad^*(y_k)^* i(\omega^k)\omega^I\cdot\ell$$

where I, J are multi-indices and $0 \neq \ell \in L_\lambda$. These correspond to the formulae for δ and δ^*. The argument of W. Schmid [43, Lemmas 5 and 6] shows that

$$[\pi] \to H_\pi \otimes \{\mathcal{H}^q(\pi^*) \otimes L_\lambda\}^H$$

is a measurable assignment of Hilbert spaces on $\hat{G}_\zeta$, and that

(7.6.5a) $H_2^{0,q}(\mathcal{L}_\lambda) = \int_{\hat{G}_\zeta} H_\pi \otimes \{\mathcal{H}^q(\pi^*) \otimes L_\lambda\}^H d\pi$

as unitary G-module. In other words

(7.6.5b) $\pi_\lambda^q = \int_{\hat{G}_\zeta} \pi\cdot\dim\{\mathcal{H}^q(\pi^*) \otimes L_\lambda\}^H d\pi.$

Thus the discrete part is

(7.6.5c) ${}^0\pi_\lambda^q = \sum_{\hat{G}_{\zeta\text{-disc}}} (\dim\{\mathcal{H}^q(\pi^*) \otimes L_\lambda\}^H)\cdot\pi.$

Note that (7.6.5c) is summation over the discrete set $W_G\backslash\{\nu \in L' : e^{\nu-\rho}|_Z = \zeta\}$.

Recall that $\Theta_{\pi_\lambda^q}^{disc}$ denotes the formal sum of the characters of the irreducible (i.e. ζ-discrete) subrepresentations of π_λ^q. Define

(7.6.6a) $F_\lambda = \sum_{q\geq 0} (-1)^q \Theta_{\pi_\lambda^q}^{disc}.$

Using (7.6.5c) we have

$$F_\lambda = \sum_{q\geq 0}(-1)^q \sum_{\pi\in\hat{G}_{\zeta\text{-disc}}} \dim\{\mathcal{H}^q(\pi^*) \otimes L_\lambda\}^H \Theta_\pi$$

$$= \sum_{\pi\in\hat{G}_{\zeta\text{-disc}}} \{\sum_{q\geq 0} (-1)^q\{\dim\{\mathcal{H}^q(\pi^*) \otimes L_\lambda\}^H\}\Theta_\pi.$$

However $\dim\{\mathcal{H}^q(\pi^*) \otimes L_\lambda\}^H$ is the multiplicity of $e^{-\lambda}$ in the representation of H on $\mathcal{H}^q(\pi^*)$. Thus

(7.6.6b) $F_\lambda = \sum_{\pi \in \hat{G}_{\zeta\text{-disc}}} (\text{coefficient of } e^{-\lambda} \text{ in } f_{\pi^*})\Theta_\pi.$

Let $[\pi_\nu] \in \hat{G}_{\zeta\text{-disc}}$, i.e. let $\nu \in L'$ with $e^{\nu-\rho}|_Z = \zeta$. Then (7.6.1)

$$\Delta T_{\pi_\nu^*}\Big|_{H \cap G'} = \Delta T_{\pi_{-\nu}}\Big|_{H \cap G'} = (-1)^{q(\nu)} \sum_{w \in W_G} \det(w) e^{-w\nu}.$$

Thus Lemma 7.6.4 says

$$f_{\pi_\nu^*} = (-1)^{n+q(\nu)} \sum_{w \in W_G} \det(w) e^{\rho - w(\nu)}.$$

In particular the coefficient of $e^{-\lambda}$ in $f_{\pi_\nu^*}$ is 0 if $\lambda + \rho \notin W_G(\nu)$, is $(-1)^{n+q(\nu)}\det(w)$ if $w(\nu) = \lambda + \rho$ for some $w \in W_G$. Now (7.6.6b) reduces to

(7.6.6c) $F_\lambda = (-1)^{n+q(\lambda+\rho)}\Theta_{\pi_{\lambda+\rho}}.$

In view of Lemma 7.5.4, the alternating sum formula (7.2.4) is equivalent to (7.6.6). This proves part 1 of Theorem 7.2.3.

7.7. We complete the proof of Theorem 7.2.3.

The vanishing statement, part 2 of Theorem 7.2.3, is proved by Griffiths and Schmid [9, Theorem 7.8] in the case where G is a connected semisimple group with finite center. Their proof goes through without change in our case.

Now suppose $H_2^{0,q}(\mathcal{L}_\lambda) = 0$ whenever $q \neq q_0$. The alternating sum formula (7.2.4), and linear independence of the Θ_π for $[\pi] \in \hat{G}_{\zeta\text{-disc}}$, then say that

(7.7.1) $[\pi_{\lambda+\rho}]$ is the discrete part $[{}^0\pi_\lambda^{q_0}]$ of $[\pi_\lambda^{q_0}]$.

In view of (7.6.1c), Corollary 5.7.2 for $\bar{\zeta}$ says that

$\{[\pi] \in \hat{G}_\zeta - \hat{G}_{\zeta\text{-disc}} : T_{\pi^*}\big|_{K \cap G'} \neq 0\}$ has Plancherel measure zero in $\hat{G}_\zeta$. Lemma 7.6.4 says $f_{\pi^*} = 0$ for almost all $[\pi] \in \hat{G}_\zeta - \hat{G}_{\zeta\text{-disc}}$. If $q \neq q_0$ then (7.6.5b) and $H_2^{0,q}(\mathcal{L}_\lambda) = 0$ force $\{\mathcal{H}^q(\pi^*) \otimes L_\lambda\}^H = 0$, so $e^{-\lambda}$ has multiplicity 0 in the representation of H on $\mathcal{H}^q(\pi^*)$. If $f_{\pi^*} = 0$ then also $e^{-\lambda}$ has multiplicity 0 in the representation of H on $\mathcal{H}^{q_0}(\pi^*)$, so $\{\mathcal{H}^{q_0}(\pi^*) \otimes L_\lambda\}^H = 0$. In summary

(7.7.2) $\{\mathcal{H}^{q_0}(\pi^*) \otimes L_\lambda\}^H = 0$ for almost all $[\pi] \in \hat{G}_\zeta - \hat{G}_{\zeta\text{-disc}}$.

The measure $\dim\{\mathcal{H}^{q_0}(\pi^*) \otimes L_\lambda\}^H d\pi$ on $\hat{G}_\zeta$ is concentrated on $\hat{G}_{\zeta\text{-disc}}$ by (7.7.2). Now (7.6.5) says $\pi_\lambda^{q_0} = {}^0\pi_\lambda^{q_0}$, so (7.7.1) tells us $[\pi_\lambda^{q_0}] = [\pi_{\lambda+\rho}]$.

Theorem 7.2.3 is proved.

§8. Measurable Orbits and Nondegenerate Series.

Let G be a reductive Lie group from the class of §3.1. If $\zeta \in \hat{Z}$ then Theorem 5.1.6 shows that Plancherel measure on $\hat{G}_\zeta$ is supported by the constituents of H-series classes that transform by ζ, as H runs over the conjugacy classes of Cartan subgroups of G. Here we work out geometric realizations for all these H-series classes. Our method is a reduction to the special case of the relative discrete series (H/Z compact), which we studied in §7.

Fix a Cartan subgroup $H = T \times A$ in G and an associated cuspidal parabolic subgroup $P = MAN$ of G. We work over measurable orbits

$$Y = G(x) \subset X = \bar{G}_C/Q \quad \text{complex flag}$$

such that (i) the G-normalizer $N_{[x]}$ of the holomorphic arc component $S_{[x]}$ is open in P and (ii) $U = \{m \in M: m(x) = x\}$ is compact modulo Z. In §8.1 we first check that G has isotropy subgroup UAN at x, $U = Z_M(M^0)U^0$ and $N_{[x]} = M^\dagger AN$. If $[\mu] \in \hat{U}$ and $\sigma \in \mathfrak{a}^*$ we show that the G-homogeneous complex vector bundle

$$p: \mathcal{V}_{\mu,\sigma} \to G/UAN = Y \quad \text{associated to } \mu \otimes e^{\rho_{\mathfrak{a}}+i\sigma}$$

is holomorphic over each holomorphic arc component of Y, in an essentially unique manner. Let K be the fixed point set of a Cartan involution that preserves H. Since μ is unitary we get a K-invariant hermitian metric on $\mathcal{V}_{\mu,\sigma}$. Since U/Z is compact we get a K-invariant assignment of hermitian metrics to the holomorphic arc components of Y. This results in Hilbert spaces $H_2^{p,q}(\mathcal{V}_{\mu,\sigma})$ of "square integrable partially harmonic (p,q)-forms" on Y with values in $\mathcal{V}_{\mu,\sigma}$: measurable ω such that (i) $\omega|_{S_{[kx]}}$ is a harmonic (p,q)-form on $S_{[kx]}$ with values in $\mathcal{V}_{\mu,\sigma}|_{S_{[kx]}}$ and L_2-norm

$\|\omega|_{S_{[kx]}}\| < \infty$ a.e. $k \in K$ and (ii) $\int_{K/Z} \|\omega|_{S_{[kx]}}\|^2 d(kZ) < \infty$. We end §8.1 by showing that the natural action of G on $H_2^{p,q}(\mathcal{V}_{\mu,\sigma})$ is a unitary representation.

The representation of G on $H_2^{0,q}(\mathcal{V}_{\mu,\sigma})$ is denoted $\pi_{\mu,\sigma}^q$. Let η_μ^q denote the representation of $M^\dagger$ on $H_2^{0,q}(\mathcal{V}_{\mu,\sigma}|_{S_{[x]}})$; we studied these in §7. Now we have a representation

$$\eta_{\mu,\sigma}^q(man) = e^{i\sigma}(a)\eta_\mu^q(m) \quad \text{on} \quad M^\dagger AN = N_{[x]}.$$

In §8.2 we prove $[\pi_{\mu,\sigma}^q] = [\mathrm{Ind}_{N_{[x]}\uparrow G}(\eta_{\mu,\sigma}^q)]$.

Our main result is Theorem 8.3.4. Split $[\mu] = [\chi \otimes \mu^0]$ where $[\chi] \in Z_M(M^0)$ and $[\mu^0] \in \hat{U}^0$ has highest weight ν such that $\nu + \rho_{\mathfrak{t}}$ is $\mathfrak{m}$-regular. Then the H-series constituents of $\pi_{\mu,\sigma}^q$ are just its irreducible subrepresentations. Their sum ${}^H\pi_{\mu,\sigma}^q$ has distribution character $\Theta^H_{\pi_{\mu,\sigma}^q}$. Further

$$\sum_{q\geqslant 0} (-1)^q\, \Theta^H_{\pi_{\mu,\sigma}^q} = (-1)^{|\Sigma_{\mathfrak{t}}^+|+q_M(\nu+\rho_{\mathfrak{t}})}\, \Theta_{\pi_{\chi,\nu+\rho_{\mathfrak{t}},\sigma}}.$$

Also, there is a constant $b_H \geqslant 0$ depending only on $[\mathfrak{m},\mathfrak{m}]$ with the property: if $\langle \nu + \rho_{\mathfrak{t}}, \phi\rangle > b_H$ for all $\phi \in \Sigma_{\mathfrak{t}}^+$ and if $q \neq q_M(\nu+\rho_{\mathfrak{t}})$ then $H_2^{0,q}(\mathcal{V}_{\mu,\sigma}) = 0$. When this vanishing holds, it combines with the alternating sum formula and some consequences of the Plancherel Theorem, yielding

$$[\pi_{\mu,\sigma}^{q_M(\nu+\rho_{\mathfrak{t}})}] = [\pi_{\chi,\nu+\rho_{\mathfrak{t}},\sigma}], \quad \text{H-series class.}$$

The proof is a matter of applying the results of §7 over every holomorphic arc component of Y, combining those results by means of the induced representation theorem of §8.2.

8.1. G is a reductive Lie group from our class specified in §3.1. As explained in §6.3, now $\bar{G} = G/Z_G(G^0)$ is a linear semisimple group with complexification $\bar{G}_C = \mathrm{Int}(\mathfrak{g}_C)$, and G acts on the complex flag manifolds of $\bar{G}_C$.

For the remainder of §8 we fix

(8.1.1a) $H = T \times A$: Cartan subgroup of G and

(8.1.1b) $P = MAN$: associated cuspidal parabolic subgroup.

In order to realize the H-series of G, we work with

(8.1.2a) $X = \bar{G}_C/Q$ complex flag manifold of $\bar{G}_C$

and

(8.1.2b) $Y = G(x) \subset X$ measurable G-orbit on X,

such that the G-normalizers of the holomorphic arc components of Y in X have the property

(8.1.2c) $N_{[x]} = \{g \in G: gS_{[x]} = S_{[x]}\}$ has Lie algebra $\mathfrak{p}$.

Since the orbit $Y = G(x)$ is measurable, it is partially complex and of flag type. Thus $S_{[x]}$ is an open M^0-orbit on the sub-flag $\bar{M}_C(x)$ where $\bar{M} = M/Z_G(G^0)$. Thus AN acts trivially on $S_{[x]}$. The isotropy group $\{g \in G: g(x) = x\} = UAN$ where $T \subset U \subset M$. We require that

(8.1.2d) $U/Z_G(G^0) = \{m \in M: m(x) = x\}/Z_G(G^0)$ is compact.

The class (6.7.1) of G-orbits discussed in §6.7 is the special case of (8.1.2) in which the orbit is integrable. We obtain a number of examples of (8.1.2) from Proposition 6.7.4 and the construction (6.7.6).

8.1.3. Lemma. *Suppose* (X,x) *given as in* (8.1.2). *Then the open orbit* $M(x) \subset \bar{M}_C(x)$ *is measurable and integrable. Further* $U = Z_M(M^0)U^0$, $U \cap M^0 = U^0$, $UM^0 = M^\dagger$, *and* $M/M^\dagger$ *enumerates the topological components of* $M(x)$. *Finally* $N_{[x]} = M^\dagger AN$, *and* $G/M^\dagger G^0$ *enumerates the topological components of* $Y = G(x)$.

Remark. $G^\dagger \subset M^\dagger G^0$ in general, but one can have $G^\dagger \neq M^\dagger G^0$. For example let $G = SL(2,R) \cup \begin{pmatrix} 1 & 0 \\ 0 & -1 \end{pmatrix} \cdot SL(2,R)$ and $\mathfrak{h} = \left\{ \begin{pmatrix} a & 0 \\ 0 & -a \end{pmatrix} : a \text{ real} \right\}$. Then $M^\dagger = M = \left\{ \pm\begin{pmatrix} 1 & 0 \\ 0 & 1 \end{pmatrix}, \pm\begin{pmatrix} 1 & 0 \\ 0 & -1 \end{pmatrix} \right\}$ so $M^\dagger G^0 = G \neq G^\dagger = G^0$.

Proof. The open orbit $M(x) \subset \bar{M}_C(x) = \bar{M}_C/Q \cap \bar{M}_C$ satisfies (7.1.1). Applying Lemma 7.1.2 to it, we get the first two assertions. For the third, $N_{[x]} = UN^0_{[x]} = UM^0AN = M^\dagger AN$, and the G-normalizer of $G^0(x)$ is $UG^0 = UM^0G^0 = M^\dagger G^0$. *q.e.d.*

We now assume (X,x) fixed as in (8.1.2). Fix

(8.1.4a) $[\mu] \in \hat{U}$ and $\sigma \in \mathfrak{a}^*$ so $[\mu \otimes e^{i\sigma}] \in (U \times A)^\wedge$.

As usual, $\Sigma^+_{\mathfrak{a}}$ is the positive $\mathfrak{a}$-root system on $\mathfrak{g}$ such that $\mathfrak{n}$ is the sum of the negative $\mathfrak{a}$-root spaces, and $\rho_{\mathfrak{a}} = \frac{1}{2} \sum_{\phi \in \Sigma^+_{\mathfrak{a}}} (\dim \mathfrak{g}^\phi)\phi$, so $\mathfrak{a}$ acts on $\mathfrak{n}$ with trace -2ρ. Now UAN acts on the representation space V_μ of μ by

(8.1.4b) $\gamma_{\mu,\sigma}(uan) = e^{\rho_{\mathfrak{a}}+i\sigma}(a)\mu(u)$.

That specifies the associated G-homogeneous complex vector bundle

(8.1.4c) $p: \mathcal{V}_{\mu,\sigma} \to G/UAN = G(x) = Y$.

8.1.5. Lemma. *There is a unique assignment of complex structures to the pieces* $p^{-1}S_{[gx]}$ *of* $\mathcal{V}_{\mu,\sigma}$ *over the holomorphic arc components of* Y, *such that each* $\mathcal{V}_{\mu,\sigma}|S_{[gx]} \to S_{[gx]}$ *is an* $N_{[gx]}$*-homogeneous holomorphic vector bundle. The assignment is* G*-invariant and is real analytic as tangent space distribution on* $\mathcal{V}_{\mu,\sigma}$.

Proof. Lemma 7.1.4 says that $p^{-1}S_{[gx]}$ has a unique complex structure for which $\mathcal{V}_{\mu,\sigma}|S_{[gx]} \to S_{[gx]}$ is an $ad(g)M^\dagger$-homogeneous holomorphic vector bundle. Each $ad(g)(an)$ is trivial on $S_{[gx]} = gS_{[x]}$ and multiplies all the fibres of $\mathcal{V}_{\mu,\sigma}|S_{[x]}$ by the same scalar $e^{\rho_{\mathfrak{a}}+i\sigma}(a)$. Now the complex structure on $p^{-1}S_{[gx]}$ is invariant by the action of $ad(g)N_{[x]} = N_{[gx]}$, so $\mathcal{V}_{\mu,\sigma}|S_{[gx]} \to S_{[gx]}$ is an $N_{[gx]}$-homogeneous holomorphic vector bundle. Finally, the assignment of complex structures to the $p^{-1}S_{[gx]}$ is G-invariant by uniqueness, thus also real-analytic. *q.e.d.*

If $z \in Y = G(x)$ then we have

(8.1.6a) T_z: holomorphic tangent space to $S_{[z]}$ at z.

Evidently $\{T_z\}_{z \in Y}$ is a G-invariant complex tangent space distribution on Y, and so it is real analytic. Thus

(8.1.6b) $\mathcal{T} = \bigcup_{z \in Y} T_z$

is a G-homogeneous real-analytic sub-bundle of the complexified tangent bundle of Y. Given integers p and q, the space of partially smooth-(p,q)-forms on Y with values in $\mathcal{V}_{\mu,\sigma}$ is

$$(8.1.6c)\ A^{p,q}(\mathcal{V}_{\mu,\sigma}): \begin{cases} \text{measurable sections } \alpha \text{ of } \mathcal{V}_{\mu,\sigma} \otimes \Lambda^p\mathcal{T}^* \otimes \Lambda^q\bar{\mathcal{T}}^* \\ \text{where } \alpha \text{ is } C^\infty \text{ on each holomorphic arc component of } Y. \end{cases}$$

If $\alpha \in A^{p,q}(\mathcal{V}_{\mu,\sigma})$ and $z \in Y$, then $\alpha|_{S_{[z]}}$ is a smooth (p,q)-form on $S_{[z]}$ with values in $\mathcal{V}_{\mu,\sigma}|S_{[z]}$, in the ordinary sense. Furthermore the $\bar{\partial}$-operator of X specifies operators $\bar{\partial}: A^{p,q}(\mathcal{V}_{\mu,\sigma}) \to A^{p,q+1}(\mathcal{V}_{\mu,\sigma})$.

We need hermitian metrics for the harmonic theory. Let θ be a Cartan involution of G with $\theta(H) = H$ and denote $K = \{g \in G: \theta(g) = g\}$. In view of (8.1.2d), $K \cap N_{[x]} = K \cap M^\dagger$ can be assumed to contain U, and we have an $M^\dagger$-invariant hermitian metric on the complex manifold $S_{[x]}$. Every holomorphic arc component of G(x) is an $S_{[kx]}$, $k \in K$; give $S_{[kx]}$ the hermitian metric such that the $k: S_{[x]} \to S_{[kx]}$ are hermitian isometries.

In other words, we have a K-invariant hermitian metric on the fibres of the bundle $\mathcal{T} \to Y$. Similarly the unitary structure of V_μ specifies an $M^\dagger$-invariant hermitian metric on the fibres of $\mathcal{V}_{\mu,\sigma}|S_{[x]}$, and that specifies a K-invariant hermitian metric on the fibres of $\mathcal{V}_{\mu,\sigma} \to Y$. Now we have K-invariant hermitian metrics on the fibres of the bundles $\mathcal{V}_{\mu,\sigma} \otimes \Lambda^p \mathcal{T}^* \otimes \Lambda^q \bar{\mathcal{T}}^* \to Y$. As in (7.1.5), that specifies Hodge-Kodaira operators

(8.1.7a) $A^{p,q}(\mathcal{V}_{\mu,\sigma}) \xrightarrow{\#} A^{n-p,n-q}(\mathcal{V}^*_{\mu,\sigma}) \xrightarrow{\tilde{\#}} A^{p,q}(\mathcal{V}_{\mu,\sigma})$

where $n = \dim_{\mathbb{C}} S_{[x]}$. It also specifies a pre-Hilbert space

(8.1.7b) $A_2^{p,q}(\mathcal{V}_{\mu,\sigma}) = \{\alpha \in A^{p,q}(\mathcal{V}_{\mu,\sigma}): \int_{K/Z} (\int_{S_{[kx]}} \alpha \wedge \# \alpha) d(kZ) < \infty\}$

whose inner product is

(8.1.7c) $\langle \alpha, \beta \rangle = \int_{K/Z} (\int_{S_{[kx]}} \alpha \wedge \# \beta) d(kZ).$

We define <u>square integrable partially-(p,q)-form</u> on Y with values in $\mathcal{V}_{\mu,\sigma}$ to mean an element of

(8.1.8a) $L_2^{p,q}(\mathcal{V}_{\mu,\sigma})$: Hilbert space completion of $A_2^{p,q}(\mathcal{V}_{\mu,\sigma})$.

$\bar{\partial}$ is densely defined on $L_2^{p,q}(\mathcal{V}_{\mu,\sigma})$ with formal adjoint $\bar{\partial}^* = -\tilde{\#}\bar{\partial}\#$; this follows from the corresponding standard fact (7.1.7a) over each holomorphic arc component. The analogue of the Hodge-Kodaira-Laplacian is

$$\square = (\bar{\partial} + \bar{\partial}^*)^2 = \bar{\partial}\bar{\partial}^* + \bar{\partial}^*\bar{\partial},$$

which is elliptic and essentially self-adjoint over every holomorphic arc component. Now $\square$ is essentially self adjoint on $L_2^{p,q}(\mathcal{V}_{\mu,\sigma})$. We also write $\square$ for the closure, which is unique self-adjoint extension on $L_2^{p,q}(\mathcal{V}_{\mu,\sigma})$. The kernel

(8.1.8b) $H_2^{p,q}(\mathcal{V}_{\mu,\sigma}) = \{\omega \in L_2^{p,q}(\mathcal{V}_{\mu,\sigma}): \square\omega = 0\}$

is the space of square integrable partially harmonic (p,q)-forms on Y with values in $\mathcal{V}_{\mu,\sigma}$. $H_2^{p,q}(\mathcal{V}_{\mu,\sigma})$ is the subspace of $A_2^{p,q}(\mathcal{V}_{\mu,\sigma})$ consisting of all elements ω such that $\omega|_{S_{[kx]}}$ is harmonic a.e. in K/Z. Also $H_2^{p,q}(\mathcal{V}_{\mu,\sigma})$ is a closed subspace of $L_2^{p,q}(\mathcal{V}_{\mu,\sigma})$ and one has an orthogonal direct sum decomposition

(8.1.8c) $L_2^{p,q}(\mathcal{V}_{\mu,\sigma}) = c\ell\ \bar\partial L_2^{p,q-1}(\mathcal{V}_{\mu,\sigma}) \otimes \bar\partial^* L_2^{p,q+1}(\mathcal{V}_{\mu,\sigma}) \otimes H_2^{p,q}(\mathcal{V}_{\mu,\sigma})$

by applying (7.1.9) over every holomorphic arc component.

8.1.9. Lemma. *The natural action* $[\tilde\pi_{\mu,\sigma}^{p,q}(g)\alpha](z) = g(\alpha(g^{-1}z))$ *of* G *on* $L_2^{p,q}(\mathcal{V}_{\mu,\sigma})$ *is a unitary representation.*

Proof. $\mathcal{V}_{\mu,\sigma} \otimes \Lambda^p \mathcal{T}^* \otimes \Lambda^q \bar{\mathcal{T}}^*$ has fibre $V_\mu^{p,q} = V_\mu \otimes \Lambda^p T_x^* \otimes \Lambda^q \bar T_x^*$ over x. If $\mu^{p,q}$ denotes the representation of U on $V_\mu^{p,q}$, then UAN acts on $V_\mu^{p,q}$ by

$$\gamma_{\mu,\sigma}^{p,q}(uan) = e^{\rho_{\mathfrak{a}}+i\sigma}(a)\mu^{p,q}(u) = e^{\rho_{\mathfrak{a}}}(a)\cdot{}'\gamma_{\mu,\sigma}^{p,q}(uan)$$

where ${}'\gamma_{\mu,\sigma}^{p,q} = \mu^{p,q} \otimes e^{i\sigma}$ is unitary. Since $e^{\rho_{\mathfrak{a}}}(a)$ is the square root of the determinant of uan on the real tangent space $\mathfrak{g}/(\mathfrak{u}+\mathfrak{a}+\mathfrak{n})$ to Y at x, now $\tilde\pi_{\mu,\sigma}^{p,q}$ is the unitarily induced representation $\mathrm{Ind}_{UAN\uparrow G}({}'\gamma_{\mu,\sigma}^{p,q})$.

q.e.d.

The representation $\tilde\pi_{\mu,\sigma}^{p,q}$ commutes with $\bar\partial$, hence also with $\bar\partial^*$ and $\square$, so $H_2^{p,q}(\mathcal{V}_{\mu,\sigma})$ is a closed invariant subspace for $\tilde\pi_{\mu,\sigma}^{p,q}$. Now we have

(8.1.10a) $\pi_{\mu,\sigma}^{p,q}$: unitary representation of G on $H_2^{p,q}(\mathcal{V}_{\mu,\sigma})$.

For convenience we also denote

(8.1.10b) $\pi_{\mu,\sigma}^{q}$: unitary representation of G on $H_2^{0,q}(\mathcal{V}_{\mu,\sigma})$.

The program of §8 is the representation of the H-series of unitary representation classes of G by the various $\pi_{\mu,\sigma}^{q}$.

8.2. We set up $\pi^q_{\mu,\sigma}$ as an induced representation from $N_{[x]} = M^\dagger AN$. Denote

(8.2.1a) $\mathcal{V}_\mu = \mathcal{V}_{\mu,\sigma}|_{S_{[x]}} \to S_{[x]}$.

It is the $M^\dagger$-homogeneous hermitian holomorphic vector bundle defined by $[\mu] \in \hat{U}$ as in Lemma 7.1.4. Thus we have

(8.2.1b) η^q_μ: unitary representation of $M^\dagger$ on $H^{0,q}_2(\mathcal{V}_\mu)$.

Now the formula

(8.2.1c) $\eta^q_{\mu,\sigma}(man) = e^{i\sigma}(a)\eta^q_\mu(m)$

defines a unitary representation of $N_{[x]} = M^\dagger AN$ on $H^{0,q}_2(\mathcal{V}_\mu)$.

8.2.2. Theorem. $[\pi^q_{\mu,\sigma}] = [\mathrm{Ind}_{N_{[x]}\uparrow G}(\eta^q_{\mu,\sigma})]$.

Proof. Let $\tilde{\pi} = \tilde{\pi}^{0,q}_{\mu,\sigma}$, the representation of G on $L^{0,q}_2(\mathcal{V}_{\mu,\sigma})$. Let $'\gamma$ denote the representation of UAN on $V_\mu \otimes \Lambda^q \bar{T}^*_x$; it is the $'\gamma^{0,q}_{\mu,\sigma} = \mu^{0,q} \otimes e^{i\sigma}$ from the proof of Lemma 8.1.9. That Lemma was proved (in case $p = 0$) by showing $[\tilde{\pi}] = [\mathrm{Ind}_{UAN\uparrow G}('\gamma)]$.

Let $\tilde{\eta}$ denote the representation of $M^\dagger$ on $L^{0,q}_2(\mathcal{V}_\mu)$, and $'\eta$ the representation of $M^\dagger AN$ there given by $'\eta(man) = e^{i\sigma}(a)\tilde{\eta}(m)$. Then $[\tilde{\eta}] = [\mathrm{Ind}_{U\uparrow M^\dagger}(\mu^{0,q})]$ and so $['\eta] = [\mathrm{Ind}_{UAN\uparrow M^\dagger AN}('\gamma)]$.

Induction by stages now says that $\tilde{\pi}$ is unitarily equivalent to $\mathrm{Ind}_{M^\dagger AN\uparrow G}('\eta)$. We need the equivalence. Let f be in the representation space of $\mathrm{Ind}_{M^\dagger AN\uparrow G}('\eta)$. In other words, f is a Borel-measurable function $G \to L^{0,q}_2(\mathcal{V}_\mu)$ such that

$$f(gman) = e^{-\rho_{\mathfrak{a}}}(a)\cdot{}'\eta(man)^{-1}f(g) \quad \text{for } g \in G,\ man \in M^\dagger AN$$

and

$$\int_{K/Z} \|f(k)\|^2 d(kZ) < \infty.$$

For almost all $g \in G$ we may view $f(g) \in L_2^{0,q}(\mathcal{V}_\mu)$ as a Borel-measurable function $M^\dagger AN \to V_\mu^q = V_\mu \otimes \Lambda^q \bar{T}_x^*$ such that

$$f(g)(puan) = {}'\gamma(uan)^{-1}[f(g)(p)] \quad \text{for} \quad p \in M^\dagger AN, \quad uan \in UAN$$

and

$$\int_{M^\dagger/U} \|f(g)(m)\|^2 d(mU) < \infty.$$

Now define

(8.2.3) $F = \Gamma(f)\colon G \to V_\mu^q$ by $F(g) = f(g)(1)$.

Then F is Borel-measurable. Using ${}'\eta = \mathrm{Ind}_{UAN \uparrow M^\dagger AN}({}'\gamma)$ we compute

$$F(guan) = f(guan)(1) = [e^{-\rho_{\mathfrak{a}}}(a)\cdot{}'\eta(uan)^{-1}f(g)](1)$$

$$= e^{-\rho_{\mathfrak{a}}}(a)\cdot{}'\gamma(uan)^{-1}\{f(g)(1)\} = e^{-\rho_{\mathfrak{a}}}(a)\cdot{}'\gamma(uan)^{-1}F(g)$$

and

$$\int_{K/Z}\left\{\int_{M^\dagger/U} \|F(km)\|^2 d(mU)\right\} d(kZ) = \int_{K/Z}\left\{\int_{M^\dagger/U} \|f(k)(m)\|^2 d(mU)\right\} d(kZ) < \infty.$$

Thus $f \to \Gamma(f) = F$ is the desired equivalence.

In the above construction, f is in the representation space of $\mathrm{Ind}_{N_{[x]}\uparrow G}(\eta^q_{\mu,\sigma})$ precisely when every $f(g)$ is annihilated by the Hodge-Kodaira-Laplace operator of $\mathcal{V}_\mu \to S_{[x]}$. The latter is equivalent to $\Gamma(f)$ being in the kernel of $\square$, the analog of the Hodge-Kodaira-Laplace operator for $\mathcal{V}_{\mu,\sigma} \to Y$. Thus the equivalence Γ of (8.2.3) restricts to the equivalence asserted in the Theorem. <u>*q.e.d.*</u>

<u>8.3</u>. We now come to the geometric realization of the H-series of unitary representation classes of G.

We construct a certain positive $\mathfrak{h}_C$-root system Σ^+ on $\mathfrak{g}_C$. First, the choice (8.1.1b) of P is a choice $\Sigma^+_{\mathfrak{a}}$ of positive $\mathfrak{a}$-root system on $\mathfrak{g}$.

Now let $\bar{M} = M/Z_G(G^0)$ and choose a system $\Pi_{\mathfrak{t}}$ of simple $\mathfrak{t}_C$-roots on $\mathfrak{m}_C$, such that the parabolic subalgebra $\mathfrak{q}_{\mathfrak{t}} \cap \bar{\mathfrak{m}}_C$ of $\bar{\mathfrak{m}}_C$ is specified (6.1.4) by $\mathfrak{t}$ and a subset $\Phi_{\mathfrak{t}} \subset \Pi_{\mathfrak{t}}$. Let $\Sigma_{\mathfrak{t}}^+$ be the corresponding positive $\mathfrak{t}_C$-root system on $\mathfrak{m}_C$. Now Σ^+ is the positive $\mathfrak{h}_C$-root system of $\mathfrak{g}_C$ that specifies $\Sigma_{\mathfrak{a}}^+$ and $\Sigma_{\mathfrak{t}}^+$ as in Lemma 4.1.7.

Let Π be the simple $\mathfrak{h}_C$-root system of $\mathfrak{g}_C$ for Σ^+. As in (6.7.6), it follows from Proposition 6.7.4 that the measurable orbit $Y = G(x) \subset X$ is integrable precisely when $\mathfrak{q}_{\mathfrak{t}} = \mathfrak{q}_{\mathfrak{t}\Phi}$ with $\Phi = \Phi_{\mathfrak{t}} \cup (\Pi - \Pi_{\mathfrak{t}})$.

As usual, denote

(8.3.1) $\rho_{\mathfrak{t}} = \frac{1}{2}\sum_{\phi\in\Sigma_{\mathfrak{t}}^+} \phi, \quad \Delta_{M,T} = \prod_{\phi\in\Sigma_{\mathfrak{t}}^+} (e^{\phi/2} - e^{-\phi/2}), \quad \tilde{\omega}_{\mathfrak{t}}(\nu) = \prod_{\phi\in\Sigma_{\mathfrak{t}}^+} \langle \nu,\phi \rangle.$

Replacing G by a Z_2-extension if necessary, Lemma 4.3.6 shows that we may assume $e^{\rho_{\mathfrak{t}}}$ and $\Delta_{M,T}$ well defined on T. Denote

(8.3.2) $L_{\mathfrak{t}} = \{\nu \in i\mathfrak{t}^*: e^{\nu}$ defined on $T^0\}$ and $L_{\mathfrak{t}}'' = \{\nu \in L_{\mathfrak{t}}: \tilde{\omega}_{\mathfrak{t}}(\nu) \neq 0\}$.

Then $\rho_{\mathfrak{t}} \in L_{\mathfrak{t}}''$. If $\nu \in L_{\mathfrak{t}}''$ then $\Sigma_{\mathfrak{t}}^+$ specifies an integer $q_M(\nu) \geqslant 0$ as in (7.2.2).

Lemma 8.1.3 gave us $U = Z_M(M^0)U^0$ and $U \cap M^0 = U^0$. In particular,

(8.3.3) $\hat{U} = \{[\chi \otimes \mu^0]: [\chi] \in Z_M(M^0)^{\wedge}$ consistent with $[\mu^0] \in \hat{U}^0\}$.

Recall the statement of Theorem 4.4.4. Retain the notation of §§8.1 and 8.2. Here is our main result.

8.3.4. Theorem. *Let* $[\mu] \in \hat{U}$, *say* $[\mu] = [\chi \otimes \mu^0]$ *as in* (8.3.3). *Let* ν *be the highest weight of* μ^0 *in the positive* $\mathfrak{t}_C$-*root system* $\Sigma_{\mathfrak{t}}^+ \cap \Phi_{\mathfrak{t}}^r$ *of* $\mathfrak{u}_C$. *Then* $\nu \in L_{\mathfrak{t}}$ *and* $\mu \in \hat{U}_{\zeta}$ *where* $\zeta \in \hat{Z}$ *coincides with* e^{ν} *on* $Z \cap M^0$ *and where* $[\chi] \in Z_M(M^0)_{\zeta}^{\wedge}$. *Now suppose* $\nu + \rho_{\mathfrak{t}} \in L_{\mathfrak{t}}''$ *and let* $\sigma \in \mathfrak{a}^*$

1. $\pi_{\mu,\sigma}^q = \sum^J \pi_{\mu,\sigma}^q$ *where* J *runs over the* G-*conjugacy classes of*

Cartan subgroups of MA. ${}^H\pi^q_{\mu,\sigma}$ *is a discrete direct sum of* H-*series representations* $\pi_{\chi,\nu',\sigma}$ *with* $\nu' \in L''_{\mathfrak{t}}$. *The other* ${}^J\pi^q_{\mu,\sigma}$ *are continuous sums of irreducible constituents of* J-*series representations of* G, *and they have no irreducible subrepresentations.*

2. ${}^H\pi^q_{\mu,\sigma}$ *has well defined distribution character* $\Theta^H_{\pi^q_{\mu,\sigma}}$, *and*

$$(8.3.5)\qquad \sum_{q\geqslant 0} (-1)^q \Theta^H_{\pi^q_{\mu,\sigma}} = (-1)^{|\Sigma^+_{\mathfrak{t}}|+q_M(\nu+\rho_{\mathfrak{t}})}\Theta_{\pi_{\chi,\nu+\rho_{\mathfrak{t}},\sigma}} .$$

3. *There is a constant* $b_H \geqslant 0$ *depending only on* $[\mathfrak{m},\mathfrak{m}]$, *with the following property. If* $|\langle \nu + \rho_{\mathfrak{t}},\phi\rangle| > b_H$ *for all* $\phi \in \Sigma^+_{\mathfrak{t}}$, *and if* $q \neq q_M(\nu + \rho_{\mathfrak{t}})$, *then* $H^{0,q}_2(\mathcal{V}_{\mu,\sigma}) = 0$.

4. *If* $H^{0,q}_2(\mathcal{V}_{\mu,\sigma}) = 0$ *whenever* $q \neq q_0$, *then* $[\pi^{q_0}_{\mu,\sigma}]$ *is the* H-*series class* $[\pi_{\chi,\nu+\rho_{\mathfrak{t}},\sigma}]$.

Remark. A generalization of the "Langlands Conjecture" to the relative discrete series of M^0, would imply $H^{0,q}_2(\mathcal{V}_{\mu,\sigma}) = 0$ for $q \neq q_M(\nu + \rho_{\mathfrak{t}})$ and $\pi^{q_M(\nu+\rho_{\mathfrak{t}})}_{\mu,\sigma} \in [\pi_{\chi,\nu+\rho_{\mathfrak{t}},\sigma}]$. In particular we conjecture that each $[\pi^q_{\mu,\sigma}]$ is a discrete sum of H-series classes.

Proof. Since $H^{0,q}_2(\mathcal{V}_\mu) \subset L^{0,q}_2(\mathcal{V}_\mu) \subset L_2(M^\dagger/Z,\zeta) \otimes (V_\mu \otimes \Lambda^q \bar{T}^*_x)$, the representation η^q_μ of $M^\dagger$ on $H^{0,q}_2(\mathcal{V}_\mu)$ is a subrepresentation of a finite multiple of the left regular representation on $L_2(M^\dagger/Z,\zeta)$. Let J_M run over the conjugacy classes of Cartan subgroups of M and $J = J_M \times A$. According to Theorem 5.1.6 now there are measures ${}^J\beta$ on $(M^\dagger)^\wedge_\zeta$, absolutely continuous with respect to Plancherel measure and concentrated on the constituents of J_M-series representation classes, and a multiplicity function m_η on $(M^\dagger)^\wedge_\zeta$, such that

(8.3.6a) $[\eta^q_\mu] = \sum [{}^J\eta^q_\mu]$ where $[{}^J\eta^q_\mu] = \int_{(M^\dagger)^\wedge_\zeta} m_\eta[\eta]d^J\beta[\eta]$.

Since $[\eta^q_\mu]$ is a subrepresentation of a finite multiple of the regular representation on $L_2(M^\dagger/Z,\zeta)$ and ${}^H\eta^q_\mu$ is the relative discrete series part,

(8.3.6b) $[{}^H\eta^q_\mu] = \sum_{(M^\dagger)^\wedge_{\zeta\text{-disc}}} m_\eta[\eta]$ discrete direct sum,

$[{}^H\eta^q_\mu]$ has well defined distribution character

(8.3.6c) $\Psi_{[{}^H\eta^q_\mu]} = \sum_{(M^\dagger)^\wedge_{\zeta\text{-disc}}} m_\eta\Psi_\eta$, and

(8.3.6d) no other ${}^J\eta^q_\mu$ has an irreducible subrepresentation.

In (8.3.6a), take tensor product with $e^{i\sigma}$ and extend to $N_{[x]} = M^\dagger AN$ by triviality on N. Then Theorem 8.2.2 gives

(8.3.7a) $[\pi^q_{\mu,\sigma}] = \sum [{}^J\psi]$ where $[{}^J\psi] = \int_{(M^\dagger)^\wedge_\zeta} m_\eta[\mathrm{Ind}_{N_{[x]}\uparrow G}(\eta \otimes e^{i\sigma})]d^J\beta[\eta]$.

Since $|M/M^\dagger| < \infty$, Theorems 4.3.8 (1,2) and 4.4.6 show that $\mathrm{Ind}_{N_{[x]}\uparrow G}(\eta \otimes e^{i\sigma})$ in ${}^J\psi$ is a finite sum of J-series representation class constituents of G. Writing that out, and combining those $J = J_M \times A$ that are G-conjugate, we get a multiplicity function n_π on $\hat{G}_\zeta$ and absolutely continuous measures ${}^J\gamma$ on $\hat{G}_\zeta$ concentrated on the constituents of the J-series classes, such that

(8.3.7b) $[\pi^q_{\mu,\sigma}] = \sum [{}^J\pi^q_{\mu,\sigma}]$ where $[{}^J\pi^q_{\mu,\sigma}] = \int_{\hat{G}_\zeta} n_\pi[\pi]d^J\gamma[\eta]$.

The correspondence between (8.3.7a) and (8.3.7b) is finite to finite. Now (8.3.6b) and (8.3.6d) tell us that $[{}^H\pi^q_{\mu,\sigma}]$ is a discrete direct sum of H-series classes $[\pi_{\chi,\nu',\sigma}]$, and that no other ${}^J\pi^q_{\mu,\sigma}$ has an irreducible subrepresentation. Part 1 of Theorem 8.3.4 is proved.

Express $[{}^H\pi^q_{\mu,\sigma}] = \sum_{\hat{G}_\zeta} r_\pi[\pi]$ where the multiplicity function r_π vanishes except on constituents of H-series classes. Using $|M/M^\dagger| < \infty$, we combine (8.3.6c) with the distribution character formula of Theorem 4.3.8. It follows that $[{}^H\pi^q_{\mu,\sigma}]$ has well defined distribution character $\Theta_{{}^H\pi^q_{\mu,\sigma}} = \sum_{\hat{G}_\zeta} r_\pi \Theta_\pi$.

Let ξ^q_μ denote the representation of M on $H_2^{0,q}(\mathcal{V}_{\mu,\sigma}|_{M(x)})$. The proof of Lemma 7.1.3 shows $\xi^q_\mu = \mathrm{Ind}_{M^\dagger\uparrow M}(\eta^q_\mu)$. Now Theorem 8.2.2 and induction by stages give

(8.3.8a) $[\pi^q_{\mu,\sigma}] = [\mathrm{Ind}_{P\uparrow G}(\xi^q_\mu \otimes e^{i\sigma})]$.

Denote the relative discrete (T-series) part of ξ^q_μ by ${}^0\xi^q_\mu$; it has character $\Psi^{disc}_{\xi^q_\mu}$. Part 1 of Theorem 8.3.4 gives

(8.3.8b) $[{}^H\pi^q_{\mu,\sigma}] = [\mathrm{Ind}_{P\uparrow G}({}^0\xi^q_\mu \otimes e^{i\sigma})]$.

The alternating sum formula (7.2.4) for M says

(8.3.8c) $$\sum_{q\geqslant 0} (-1)^q \Psi^{disc}_{\xi^q_\mu} = (-1)^{|\Sigma^+_{\mathfrak{t}}|+q_M(\nu+\rho_{\mathfrak{t}})} \Psi_{\eta_{\chi,\nu+\rho_{\mathfrak{t}}}}.$$

The formulae of Theorem 4.3.8 are valid for $\eta = {}^0\xi^q_\mu$ and for $\eta = \eta_{\chi,\nu+\rho_{\mathfrak{t}}}$.

Now the alternating sum formula (8.3.5) follows directly from (8.3.8).

Let $b_H \geqslant 0$ denote the constant in Part 2 of Theorem 7.2.3 for M. It depends only on $[\mathfrak{m},\mathfrak{m}]$. Suppose $|\langle \nu + \rho_{\mathfrak{t}}, \phi\rangle| > b_H$ for every $\phi \in \Sigma^+_{\mathfrak{t}}$. Then $H_2^{0,q}(\mathcal{V}_{\mu,\sigma}|_{M(x)}) = 0$, by Theorem 7.2.3 (2), so (8.3.8a) tells us $H_2^{0,q}(\mathcal{V}_{\mu,\sigma}) = 0$.

The last assertion of Theorem 8.3.4 is immediate from the alternating sum formula and Theorem 7.2.3 (3). *q.e.d.*

8.4. Theorem 8.3.4 provides explicit geometric realization for "most" of the H-series classes of unitary representation of G. It provides implicit geometric realizations for all the H-series classes. Theorem 7.2.3 is the special case where H/Z is compact. In view of the Plancherel Theorem 5.1.6, we now have, for every $\zeta \in \hat{Z}$, geometric realizations for a subset of $\hat{G}_\zeta$ that supports Plancherel measure there.

References

1. A. Andreotti and E. Vesentini, *Carleman estimates for the Laplace-Beltrami operator on complex manifolds*, Inst. Hautes Études Sci. Publ. Math. 25 (1965), 81-130.

2. L. Auslander and C. C. Moore, "Unitary Representations of Solvable Lie Groups". Amer. Math. Soc. Memoir 62. Providence, 1966.

3. V. Bargmann, *On unitary ray representations of continuous groups*, Annals of Math. 59 (1954), 1-46.

4. R. Bott, *Homogeneous vector bundles*, Annals of Math. 66 (1957), 203-248.

5. F. Bruhat, *Sur les représentations induites des groupes de Lie*, Bull. Soc. Math. de France 84 (1956), 97-205.

6. J. Dixmier, "Les C*-algèbres et leurs Représentations". Gauthier-Villars, Paris, 1964.

7. M. Duflo, *Sur les extensions des représentations irréductibles des groupes de Lie nilpotents*, Ann. Sci. Éc. Norm. Sup.(4) 5 (1972), 71-120.

8. R. Godement, *Sur les relations d'orthogonalité de V. Bargmann*, C. R. Acad. Sci. Paris 225 (1947), 521-523 and 657-659.

9. P. A. Griffiths and W. Schmid, *Locally homogeneous complex manifolds*, Acta Math. 123 (1969), 253-302.

10. S. Grosser and M. Moskowitz, *On central topological groups*, Trans. Amer. Math. Soc. 127 (1967), 317-340.

11. ________, *Representation theory of central topological groups*, Trans. Amer. Math. Soc. 129 (1967), 361-390.

12. ________, *Harmonic analysis on central topological groups*, Trans. Amer. Math. Soc. 156 (1971), 419-454.

13. Harish-Chandra, *Representations of a semisimple Lie group on a Banach space*, I, Trans. Amer. Math. Soc. 75 (1953), 185-243.

14. ________, *Representations of semisimple Lie groups*, II, Trans. Amer. Math. Soc. 76 (1954), 26-65.

15. ________, *Representations of semisimple Lie groups*, III, Trans. Amer. Math. Soc. 76 (1954), 234-253.

16. Harish-Chandra, *The Plancherel formula for complex semisimple Lie groups*, Trans. Amer. Math. Soc. 76 (1954), 485-528.

17. ________, *Representations of semisimple Lie groups*, IV, Amer. J. Math. 77 (1955), 743-777.

18. ________, *Representations of semisimple Lie groups*, V, Amer. J. Math. 78 (1956), 1-41.

19. ________, *Representations of semisimple Lie groups*, VI, Amer. J. Math. 78 (1956) 564-628.

20. ________, *The characters of semisimple Lie groups*, Trans. Amer. Math. Soc. 83 (1956), 98-163.

21. ________, *Invariant eigendistributions on a semisimple Lie group*, Trans. Amer. Math. Soc. 119 (1965), 457-508.

22. ________, *Two theorems on semisimple Lie groups*, Annals of Math. 83 (1966), 74-128.

23. ________, *Discrete series for semisimple Lie groups*, I, Acta Math. 113 (1965), 241-317.

24. ________, *Discrete series for semisimple Lie groups*, II, Acta Math. 116 (1966), 1-111.

25. ________, *Harmonic analysis on semisimple Lie groups*, Bull. Amer. Math. Soc. 76 (1970), 529-551.

26. ________, *On the theory of the Eisenstein integral*, Springer-Verlag Lecture Notes in Mathematics 266 (1971), 123-149.

27. T. Hirai, *The characters of some induced representations of semisimple Lie groups*, J. Math. Kyoto Univ. 8 (1968), 313-363.

28. R. Hotta, *On realization of the discrete series for semisimple Lie groups*, J. Math. Soc. Japan 23 (1971), 384-407.

29. B. Kostant, *Lie algebra cohomology and the generalized Borel-Weil theorem*, Annals of Math. 74 (1961), 329-387.

30. R. Lipsman, *On the characters and equivalence of continuous series representations*, J. Math. Soc. Japan 23 (1971), 452-480.

31. ________, *Representation theory of almost connected groups*, to appear.

32. G. W. Mackey, *Induced representations of locally compact groups*, II: *the Frobenius reciprocity theorem*, Annals of Math. 58 (1953), 193-221.

33. G.W. Mackey, "The Theory of Group Representations". University of Chicago lecture notes, 1955.

34. ________, *Borel structures in groups and their duals*, Trans. Amer. Math. Soc. 85 (1957), 134-165.

35. ________, *Unitary representations of group extensions*, I, Acta Math. 99 (1958), 265-311.

36. C.C. Moore, *Extensions and low dimensional cohomology theory of locally compact groups*, I, Trans. Amer. Math. Soc. 113 (1964), 40-63.

37. ________, *Extensions and low dimensional cohomology theory of locally compact groups*, II, Trans. Amer. Math. Soc. 113 (1964), 64-86.

38. M.S. Narasimhan and K. Okamoto, *An analogue of the Borel-Weil-Bott theorem for hermitian symmetric pairs of noncompact type*, Annals of Math. 91 (1970), 486-511.

39. R. Parthasarathy, *Dirac operator and the discrete series*, Annals of Math. 96 (1972), 1-30.

40. ________, *A note on the vanishing of certain* L^2*-cohomologies*, to appear.

41. M.A. Rieffel, *Square integrable representations of Hilbert algebras*, J. Functional Analysis 3 (1969), 265-300.

42. L.P. Rothschild, *Orbits in a real reductive Lie algebra*, Trans. Amer. Math. Soc. 168 (1972), 403-421.

43. W. Schmid, *On a conjecture of Langlands*, Annals of Math. 93 (1971), 1-42.

44. I.E. Segal, *An extension of Plancherel's formula to separable unimodular groups*, Annals of Math. 52 (1950), 272-292.

45. M. Sugiura, *Conjugate classes of Cartan subalgebras in a real semisimple Lie algebra*, J. Math. Soc. Japan 11 (1959), 374-434.

46. J.A. Tirao, *Square integrable representations of semisimple Lie groups*, Trans. Amer. Math. Soc., to appear.

47. J.A. Tirao and J.A. Wolf, *Homogeneous holomorphic vector bundles*, Indiana Univ. Math. J. 20 (1970), 15-31.

48. J.A. Wolf, "Spaces of Constant Curvature", 2nd edition, J.A. Wolf, Berkeley, 1972.

49. ________, *The action of a real semisimple Lie group on a complex flag manifold, I: Orbit structure and holomorphic arc components*, Bull. Amer. Math. Soc. 75 (1969), 1121-1237.

50. J.A. Wolf, *Complex manifolds and unitary representations*, Springer-Verlag Lecture Notes in Mathematics 185 (1971), 242-287.

51. ________, *Fine Structure of hermitian symmetric spaces*, "Geometry and Analysis of Symmetric Spaces", Marcel Dekker, New York, 1972.

52. ________, *The spectrum of a reductive Lie group*, Amer. Math. Soc. Proceedings of Symposia on Pure Mathematics 25 (1974), to appear.

53. ________, *Geometric realizations of representations of reductive Lie groups*, Amer. Math. Soc. Proceedings of Symposia in Pure Mathematics 25 (1974), to appear.

54. J.A. Wolf and A. Korányi, *Generalized Cayley transformations of bounded symmetric domains*, Amer. J. Math. 87 (1965), 899-939.

55. R. A. Kunze, *On the Frobenius Reciprocity theorem for square integrable representations*, Pacific J. Math., to appear.

56. A. Wawrzyńczyk, *On the Frobenius-Mautner Reciprocity theorem*, Bull. Acad. Pol. Sci. 20 (1972), 555-559.

57. J.A. Wolf, *Partially harmonic spinors and representations of reductive Lie groups*, J. Functional Analysis, to appear.

University of California at Berkeley